Population Mobility and Indigenous Peoples in Australasia and North America

For population analysts, two of the most difficult issues to grapple with are Indigenous populations and mobility. Indigenous peoples in Australia, New Zealand, Canada and the United States comprise the descendents of the original inhabitants of these lands. One impact of colonization on these peoples has been their widespread dispersion and spatial redistribution. They are now located in major cities and the remotest of localities, either within traditional homelands, or far from them. No systematic analysis exists of the geographic movement of these peoples, either historically or in contemporary times.

With contributions from leading scholars, this book draws together relevant research findings to produce the first comprehensive overview of Indigenous peoples' mobility. Chapters draw from a range of disciplinary perspectives, and from a diversity of regions and nation-states. Within nations, mobility is the key determinant of local population change, with implications for service delivery, needs assessment, and governance. Mobility also provides a key indicator of social and economic transformation. As such, it informs both social theory and policy debate.

For much of the twentieth century conventional wisdom anticipated the steady convergence of socio-demographic trends, seeing this as an inevitable concomitant of the development process. However, the patterns and trends in population movement observed in this book suggest otherwise, and provide a forceful manifestation of changing race relations in these New World settings.

John Taylor is a Senior Fellow at the Australian National University's Centre for Aboriginal Economic Policy Research. For the past twenty years his research interests have focused on the measurement of demographic and economic change among Indigenous Australians. **Martin Bell** is Senior Lecturer in Geography and Director of the Queensland Centre for Population Research at the University of Queensland. His major research interests focus on population mobility and internal migration, especially in a cross-national comparative framework.

Routledge Research in Population and Migration
Series Editors: Paul Boyle and Mike Parnwell

Population Mobility and Indigenous Peoples in Australasia and North America

Edited by John Taylor
and Martin Bell

Routledge
Taylor & Francis Group

LONDON AND NEW YORK

First published 2004
by Routledge
2 Park Square, Milton Park, Abingdon, Oxon, OX14 4RN

Simultaneously published in the USA and Canada
by Routledge
270 Madison Ave, New York NY 10016

Routledge is an imprint of the Taylor & Francis Group

Transferred to Digital Printing 2005

Typeset in Baskerville by Exe Valley Dataset Ltd, Exeter, Devon

British Library Cataloguing in Publication Data
A catalogue record for this book is available from the British Library

Library of Congress Cataloging in Publication Data
A catalog record for this book has been requested

ISBN 0–415–22430–6

Contents

Local contingency   161

8 **The politics of Māori mobility**   163

   MANUHUIA BARCHAM

   Early Māori mobility patterns 163
   Rural–urban migration 165
   Rising urban politicization 168
   Return-migration 171
   Conclusion 178

9 **American Indians and geographic mobility: some parameters
   for public policy**   184

   C. MATTHEW SNIPP

   The geographic mobility of Indigenous Americans 185
   Geographic mobility, Indigenous peoples and public policy 186
   The implications of duration and place 190
   The policy challenge 197

10 **The formation of contemporary Aboriginal settlement
    patterns in Australia: government policies and programmes**   201

    ALAN GRAY

    Control over mobility 201
    Urbanization 208
    Housing and mobility 211
    Conclusion 219

11 **Myth of the "walkabout": movement in the Aboriginal
    domain**   223

    NICOLAS PETERSON

    Aboriginal domains 224
    Beats, runs and lines 226
    Contemporary ceremonial movement 230
    Conclusion 234

12 **The social underpinnings of an "outstation movement"
    in Cape York Peninsula, Australia**   239

    BENJAMIN RICHARD SMITH

    Historical impacts on mobility patterns 240
    Colonial impacts on population mobility 243
    Decentralization and self-determination 246
    Contemporary mobility patterns 250
    Mobility, policy and the state 253

Figures

Tables

Contributors

Manuhuia Barcham is a doctoral student at the Australian National University and has recently completed a fellowship at Oriel College, University of Oxford. His research interests lie within the fields of political theory, ethics and jurisprudence and the relationship that these disciplines have with practical problems of government. He has published a number of chapters and articles on Indigenous peoples, focusing particularly on New Zealand Maori.

Daniel Beavon is the Director of the Strategic Research and Analysis Directorate at the Department of Indian and Northern Affairs Canada. He has a BA and MA in criminology from Simon Fraser University and pursued doctoral studies in sociology at Carleton University. Prior to his current position Dan worked in programme evaluation, performance measurement, and research with both the Correctional Service Canada and the Department of Justice.

Richard Bedford is Professor of Geography and Convenor of the Migration Research Group in the Population Studies Centre, Waikato University. He is a Fellow, Royal Society of NZ; Convenor, Social Sciences Sub Commission of the NZ National Commission for UNESCO; and Member, IGU Commission on Population and Environment. Major research interests centre around migration and development in the Pacific Islands and New Zealand. Several publications on population movement in Vanuatu and Fiji address mobility transitions.

Martin Bell is Senior Lecturer in Geography and Director of the Queensland Centre for Population Research at the University of Queensland. Major research interests focus on population mobility and internal migration, and on demographic forecasting and projection, especially at the local and regional level.

Stewart Clatworthy is the owner and principal researcher of Four Directions Project Consultants, a Winnipeg-based consulting firm specializing in Aboriginal studies, socioeconomic research, information systems development and programme evaluation. He has been active in Aboriginal research for over 20 years and has completed more than 80 studies focusing on Aboriginal demography and migration, population projections, and socioeconomic conditions.

Martin Cooke is a doctoral student in the Department of Sociology at the University of Western Ontario, where he is studying demography and social stratification. He has recently received a Canadian Policy Research Award Graduate Prize for his unpublished Master's thesis, 'On Leaving Home: Return and Circular Migration between First Nations and Prairie Cities', and holds a Social Sciences and Humanities Research Council Doctoral Fellowship.

Karl Eschbach is Associate Professor of Sociology at the University of Houston, and is a research associate at the University's Center for Immigration Research. Dr Eschbach's research has focused on the relationship between spatial distribution of ethnic populations and the dynamics of ethnic identity. He has published several articles that analyse changes in the ethnic identification of mixed ancestry American Indians in the twentieth century.

James S. Frideres PhD is a Professor of Sociology and Associate Vice-President (Academic) at the University of Calgary. Dr Frideres has just published the sixth edition of his book *Aboriginal Peoples in Canada* and it continues to be the major textbook in the country. His research interests include immigrant integration and identity. His awards include the outstanding teacher award at the University of Calgary and the Distinguished Service Award from the Canadian Sociology and Anthropology Association.

Alan Gray (1946–2001) was dedicated to contributing to knowledge on Aboriginal population issues. These contributions encompass the areas of Aboriginal fertility, family formation, mortality, health and human rights. He previously held positions at the Australian Bureau of Statistics, the National Centre for Epidemiology and Population Health (ANU), and the Demography Program (ANU). He was an associate of the Centre for Aboriginal Economic Policy Research, and was employed at the Institute for Population and Social Research, Mahidol University, Thailand.

Eric Guimond is a demographer with experience in research and development, presently employed by the Strategic Research and Analysis Directorate of Indian and Northern Affairs Canada. His education background includes demography, community health, physical education and Aboriginal studies. He has expertise in projection models of population and Aboriginal groups. The topic of his PhD studies is the ethnic mobility of Aboriginal populations in Canada.

Madeline A. Kalbach PhD is the holder of the Chair in Canadian Ethnic Studies at the University of Calgary where she is also a Professor of Sociology. Her research interests are multiculturalism, immigration and ethnic groups. She has published widely in these areas. She is the author of the book *Ethnic Groups and Marital Choices* (UBC Press) and co-editor and author of *Perspectives on Ethnicity in Canada* (Harcourt Brace Canada). Professor Kalbach is a member of the University of Calgary Senate.

Warren E. Kalbach PhD FRSC is Professor Emeritus, University of Toronto and Adjunct Professor, University of Calgary. Professor Kalbach is an eminent Canadian demographer. He is co-editor and author of *Perspectives on Ethnicity in Canada* (Harcourt Brace Canada), and is also a co-author of *Demographic Bases of Canadian Society* (two editions) and *Canadian Population*. He is the author of several census monographs including *The Impact of Immigration on Canada's Population* and *The Adjustment of Canada's Immigrants and their Descendants*.

Bruce Newbold is an Associate Professor of Geography at McMaster University, where he received his PhD in 1994. His research interests include internal migration, immigration, and population health. With over twenty-five refereed journal articles or book chapters published, he has received funding from the Canadian Institutes of Health Research, Social Science and Humanities Research Council of Canada, the National Science Foundation, and the Social Science Research Council.

Mary Jane Norris is a Senior Research Manager with the Strategic Research and Analysis Directorate of the Department of Indian and Northern Affairs Canada. Prior to her current position, she focused on Aboriginal research in the Demography Division of Statistics Canada. She has specialized in Aboriginal demography over the past twenty years and has published on issues related to Aboriginal migration, population projections and languages.

Nicolas Peterson is a Reader in Anthropology at the Australian National University. He has carried out field research in Arnhem Land and Central Australia with Aboriginal people. His research interests include territorial and marine tenure systems, social organization, social change and relations between Indigenous people and the state. He has recently co-edited *Citizenship and Indigenous Australians* with W. Sanders (1998 Cambridge) and *Customary Marine Tenure in Australia* with B. Rigsby (1998, Oceania Monographs).

Ian Pool is Demography Professor, Director Population Studies Centre, Waikato University , New Zealand; Fellow Royal Society of NZ; Member, Scientific Committee on Age Structure and Policy, International Union for the Scientific Study of Population; and Scientific Consultant, Committee for International Cooperation in National Demographic Research. His numerous research publications include those on population, development and policy, fertility, mortality, and Maori, on whom his most recent book is *Te Iwi Maori*

Benjamin Richard Smith is a Postdoctoral Research Fellow at the Centre for Aboriginal Economic Policy Research at the Australian National University where he is currently working on an Australian Research Council-funded research project. He has conducted academic and applied research with Aboriginal people across northern Queensland and his research interests include land rights and native title, population mobility, decentralization and social change.

C. Matthew Snipp is a Professor in the Department of Sociology at Stanford University. Before moving to Stanford in 1996, he was a Professor of Rural Sociology and Sociology, and Director of the American Indian Studies Program at the University of Wisconsin Madison. Professor Snipp has published three books and over 60 articles and book chapters on American Indian demography, economic development, poverty and unemployment. His tribal background is Oklahoma Cherokee and Choctaw.

John Taylor is a Senior Fellow at the Australian National University's Centre for Aboriginal Economic Policy Research, and a member of the Australian Population Association. For the past twenty years his major research interests have focused on the measurement of demographic and economic change among Indigenous Australians. He has published widely on these issues in Australian and international books and journals.

Acknowledgements

The germ of an idea for this book was sown at the Remote Regions sessions of the Western Regional Science Association annual conference in California as long ago as the late 1980s, and from John Taylor's subsequent research posting at the University of Alaska in 1990. For an Australian geographer, this exposure to the remote regions of North America was striking in its parallels, not least in regard to the demography of Indigenous peoples. It is appropriate that recognition be afforded to Tom Morehouse, Lee Huskey and Linda Ellanna, then of the University of Alaska, and Ken Coates, then of the University of Victoria, for illuminating the potential benefit of cross-national comparison.

Further development took place around the fringes of conference meetings, and discussions in particular with Graeme Hugo, Ron Skeldon, and Ian Pool at the Adelaide conference of the Australian Population Association helped to consolidate the project, as did a seminar at the European Institute for Population Studies at the University of Liverpool, which Bill Gould kindly facilitated.

In sketching out an Indigenous people's mobility transition with Martin Bell for the *International Journal of Population Geography* we were encouraged by comments of the editor, Huw Jones, to pursue matters further and accordingly invited papers from across the Tasman and from North America in pursuit of the present volume. In this endeavour, we have been supported from the outset by the editor of the Routledge Research in Population and Migration Series, Paul Boyle, while Jon Altman of the Centre for Aboriginal Economic Policy Research at the Australian National University (ANU) assisted greatly with a generous allocation of staff time and resources. Acknowledgement is also made of the support provided by the Queensland Centre for Population Research at the University of Queensland.

As with all such productions, the supporting cast has been considerable. Particular mention should be made of the patient assistance provided by the Routledge editorial team in London, including Simon Whitmore, Annabel Watson, and Amritpal Bangard. At the ANU, colleagues Jon

Altman, Diane Smith, Will Sanders, Frances Morphy, and Nic Peterson offered helpful advice and made useful comment on selected papers. Sally Ward provided strategic editorial assistance, while Ian Heyward of the ANU cartography unit kindly drafted many of the maps and diagrams.

1 Introduction

New World demography

John Taylor and Martin Bell

Indigenous peoples in the new world countries of Australia, New Zealand, Canada and the United States comprise those descendants of the original inhabitants of these lands, who retain cultural difference from majority settler populations that have usurped their territory, and who identify themselves as Indigenous (Taylor 2003). Aside from initial population decline, one of the most tangible and lasting demographic impacts of colonization on these peoples has been their widespread dispersion and spatial redistribution. They are now located in a wide variety of residential settings ranging from major metropolitan areas to the remotest and smallest of localities, either within their traditional homelands or far from them.

Despite widespread recognition of this impact, there has been no systematic analysis of the geographic movement of these peoples, either historically or in contemporary times. Although the literature concerned with Indigenous demography and mobility is not inconsiderable, it tends to be unsystematic, spatially restricted, and generally inaccessible to wide readership. Furthermore, what knowledge we do have of population movement among Indigenous peoples remains, all too often, a by-product of some other investigation into social and economic conditions with only few attempts to make it the primary focus of attention. What is particularly lacking, as a consequence, is a sense of the overall spatial structure of Indigenous mobility behaviour within which new studies of population movement might be situated, and research priorities set. One glaring effect of this lack of context is an inability to compare patterns of movement among Indigenous peoples with those observed at national and regional levels for the majority populations in each country, about whom much more is known. This has drawbacks, not only in regard to the development of migration theory, but also in terms of applications to policy and planning, since effective policy development depends fundamentally upon understanding the distinctiveness of Indigenous social and economic behaviour.

There are clear commonalities in the colonial and post-colonial experiences of Indigenous peoples in North America and Australasia that make them obvious candidates for comparative research. There are also several advantages to be gained from drawing together the disparate evidence on Indigenous mobility from similar cultural settings. Cross-national comparison helps reveal common patterns and trends, it assists in the exploration of shared research problems, it provides for the development of appropriate analytical methods, and contributes to the construction, and testing of theoretical frameworks. At a fundamental level, it also encourages rigour in analytical research within individual country settings. In other aspects of Indigenous demography considerable progress on this path has already been made, with a number of comparative analyses of trends in fertility and mortality (Hogg 1992; Kenen 1987; Kunitz 1990, 1994; Pool 1986; Ross and Taylor 2002). What is conspicuously lacking, to complete the picture, are equivalent comparisons of population mobility. This volume aims to provide such a synthesis and delineates, for the first time, an emerging field of population research. While each of the invited chapters is pathbreaking in its own way as an attempt to summarize available knowledge, or to focus on specific aspects of Indigenous mobility, collectively the papers lay the first foundation for cross-national comparison. The background to this initiative is outlined below.

Background

Some years ago we argued in the *International Journal of Population Geography* that Indigenous peoples in Australasia and North America exhibited demographic regimes that were quite distinct from those observed for the majority populations of which they had come to form a part (Taylor and Bell 1996). The thrust of the argument was that differences in mobility represented one dimension of an 'enclave demography' circumscribed by social and cultural constructions of identity and subject to the impact of non-Indigenous settlement, the imposition of state control, and regimes of Indigenous self-governance. Structurally, the socio-economic position of these Indigenous populations has been described as resembling that of the 'Third World in the First' (Young 1995). However, unlike populations in the Third World, the demography of Indigenous groups in Australasia and North America is as much a manifestation of inequitable power relations and marginalization in the midst of plenty, as it is to do with any lack of development *per se* (Bodley 1990; Gray 1985).

In the same paper we drew attention to the similarity of mobility patterns and trends among Indigenous peoples in Australasia and North America. The commonalities we observed included: major redistribution following colonial settlement due to displacement by expanding colonial frontiers; high levels of mobility, especially short-term circulatory movements; high regional variability in census-based measures of movement; a sustained presence in non-metropolitan areas; the rise and subsequent abatement of post-war migration to urban centres; and evidence of more recent return migration to places and regions of cultural significance.

These similarities merit some elaboration. One notable feature is that studies in all four countries report rapid post-war rural–urban migration, and that the origins are traced to much the same reasons: public policy intervention combined with the search for better employment prospects and housing. At the same time, it is interesting to note that doubts expressed in Australia as to the true contribution of migration to Indigenous urbanization have also been raised in other New World contexts. For example, census data on race, ancestry and life-time migration indicate that almost all the apparent redistribution of the American Indian population evident from successive post-war censuses was attributable to changes in ethnic identification rather than to migration (Eschbach 1993).

There is also evidence from all four countries to demonstrate that the integration of Indigenous peoples into the urban system has been less than wholesale. Thus, in Australia, most Indigenous people still live distant from the major cities, in contrast to the majority population. Similarly, in the United States and Canada, reservation populations continue to comprise a substantial share of the total American Indian population (Snipp 1989). Though they may live in urban areas for large amounts of time, Canadian Indians still retain social ties in rural areas, and in recent times net flows have been away from the cities towards rural reserves (Frideres 1983: 198; Norris 1990: 54–6). Even in New Zealand, where the migration of Indigenous people to urban areas has been most pronounced, movement of Maori in the 1970s occurred predominantly within rural areas or between town and hinterland (Poulsen *et al.* 975: 320). More recently, Pool (1991: 204–6) and Scott and Kearns (2000) report that Maori redistribution patterns have become increasingly complex due to the growing incidence of return migration to rural communities. In Alaska, too, Taylor (1991) has noted how strong economic and political forces have emerged to sustain a dispersed, rural settlement pattern among Alaska Natives. Despite clear evidence of movement towards urban centres (Hamilton and Seyfrit 1994), the more striking feature is how Native Alaskans retain connections with vibrant rural communities (Fienup-Riordan 1983; Jorgensen 1990). This recurring theme, emphasizing widespread continuity, and even rejuvenation of a rural population base, runs counter to the mainstream mobility experience in developed countries, and underscores the fact that Indigenous populations have homelands *within* modern nation-states.

So pervasive are these commonalities among the four countries that our 1996 paper advanced the proposition of a global Indigenous variant to Zelinsky's mobility transition hypothesis (Zelinsky 1971). While this may have been optimistic, it was difficult to overlook the degree of cross-national correspondence in mobility behaviour that appeared to exist between Indigenous peoples in these particular settings. At the very least, the evidence suggested that potentially fruitful lines of comparative inquiry might exist, which could help establish, more precisely, the extent and nature of any consistencies over time and space, and whether these might be linked to common underlying processes. It was against this background that we invited leading scholars of varying disciplinary persuasion, who had (almost uniquely as a group) written on issues related to Indigenous population movement, to contribute to an innovative synthesis of relevant research. The present collection of essays is the outcome, and it represents the first comprehensive overview of Indigenous peoples' mobility, at least in the New World countries of Australasia and North America.

This delineation of an emergent corpus of research addresses a growing academic and policy interest in Indigenous population issues that emerged in the latter decades of the twentieth century at both national and international levels. Within nations, mobility is now the key determinant of regional and local population change, with implications for modes of service delivery, needs assessment, and governance structures. More broadly, movement propensities and patterns of spatial redistribution also provide key indicators of social and economic transformation, marking individual and group responses to developmental and modernizing forces. As such, they inform both social theory and policy debate.

With regard to theoretical positions, it is worth recalling that for much of the twentieth century conventional wisdom anticipated the steady convergence of socio-demographic trends, seeing this as an inevitable concomitant of the development process. Convergence, as Wilson (2001) observed, lies at the heart of demographic transition theory and has provided the central underpinning for the massive body of empirical research into global population trends conducted over the past 50 years (Caldwell and Caldwell 2001; Bongaarts 2002). The strength of this belief in convergence is perhaps most powerfully exemplified in the persistent assumption in United Nations' population projections that all the world's countries would eventually tend towards replacement fertility. Only in very recent years have scholars begun to question the inevitability of such demographic homogeneity (Jones 1993; Caselli *et al.* 2002).

Expectations regarding mobility have been less strongly drawn into the convergence mould. In part, this reflects a lack of universally accepted measures for comparing mobility behaviour between nations and groups, of the type that exist for the other components of population change (Bell *et al.* 2002). Nonetheless, there are clear indications that spatial behaviour would tend towards universal norms. For example, urbanization (itself fundamentally dependent upon migration) is seen as an inevitable co-requisite of the development process (Skeldon 1997), while Zelinsky's (1971) mobility hypothesis is explicitly crafted around the notion of functional, inviolate links between mobility, modernization and the stages of demographic transition.

For Indigenous peoples of the New World, as for others wherever located, this expected convergence involved conflict between tradition and change; Indigenous culture and modernity; townsmen and tribesmen. For much of the period of post-European contact, the inevitable prospect facing Indigenous cultures was seen as assimilation into national socio-cultural and economic systems involving an ultimate loss of separate identity (Sahlins 1999). However, one of the surprises of late capitalism at the commencement of the twenty-first century is that Indigenous peoples, not least in New World countries, have flourished demographically, not by losing their culture – or more specifically their identity – but, to use Sahlin's (1999: ix) phrase, by "Indigenizing modernity". This notion, that encapsulated peoples have not only survived but bring expression to contemporary social and economic outcomes, is germane to this volume, and at least part of the project is to test the validity of this idea with respect to mobility patterns and trends within different jurisdictions.

General grounds for a specific focus on the demography of separately identified ethnic populations within developed nations have been advanced before. For example, Robinson (1992) has developed a series of arguments for separate analysis of ethnic groups in the United Kingdom, and we adapt these below for Indigenous populations:

- *Locational arguments:* despite national minority status in each of the New World countries, there is considerable regional diversity of Indigenous representation, for example, the demography of large tracts of remote Australia and North America, in particular, is effectively the demography of its Indigenous inhabitants.
- *Social justice arguments:* policy makers in all four countries are committed to achieving social justice for their Indigenous peoples and an understanding of population mobility is an essential input to this endeavour.
- *Conceptual arguments:* available evidence suggests that mobility behaviour among Indigenous peoples is not simply a sub-set of the pattern observed overall and that there are unique cultural and structural factors which produce distinct mobility outcomes.
- *Social science arguments:* given the infancy of the field, there is a need to develop comprehensive, robust measures which allow rigorous comparison of mobility and migration within and between countries.
- *Contextual arguments:* a history of widespread dispossession of land and variable modes of articulation with settler societies has created unique social and economic contexts for Indigenous mobility. While Indigenous people have been subject to redistributional influences of government policy, mobility has also been influenced by the principles and practice of Indigenous self-determination and self-management.

Together these arguments provide a compelling foundation for research into the mobility of Indigenous peoples and underscore the value of a cross-national comparative perspective.

Organization of the book

The framework adopted for this volume emanates from the basic proposition outlined above: namely, that the settlement patterns and mobility
behaviour of Indigenous minorities in the four countries of interest are
fundamentally a product of the changing relationship between Indigenous
cultures and the encapsulating state. This has wide-ranging consequences:
for the measurement of mobility (because of cultural constructions of
movement and social constructions of identity); for the intensity and
patterns of movement (because of unique relationships between Indigenous
groups and dominant institutional structures); for individual behaviour
(because mobility is heavily mediated by cultural milieux); and ultimately
for policy (because of the scale and specificity of the socio-economic
disadvantages that confront Indigenous populations). These areas of
analytical concern provide the core underlying themes of the book. In
addressing these, the book is structured into three main parts, each of
which adopts a particular perspective on the nature of Indigenous mobility.
In terms of spatial scale, these move from broad national level overviews
down to local case studies.

Part I (International Perspectives) comprises four chapters that summarize the key literature and data in each country to identify broad
national trends, themes, patterns and issues of analytical relevance. In
Chapter 2, Taylor and Bell examine the evidence for continuity and change
in Indigenous Australian population mobility. While substantial change is
identified, particularly in the form of rising urbanization, they emphasize
the importance of Indigenous agency in mobility behaviour and outcomes.
In so doing, they build upon earlier work by Hugo (1988) to elaborate an
Australian model of Indigenous mobility transition, with refinements
focused on a dilution in the strength of post-war urbanization, a more
widespread recognition of customary forms of circular movement, and a
contemporary phase of population dispersion.

Bedford and Pool are more forthright in their groundbreaking and
detailed examination of a Maori mobility transition set against Zelinsky's
hypothesized phases – the first such analysis of its kind for any Indigenous
population (Chapter 3). They identify significant cultural, historic and
structural circumstances in Aotearoa/New Zealand that produce patterns of
movement tending towards convergence with the majority non-Maori
(Pakeha) population, including a sizeable overseas migration component –
outcomes which appear to place Maori somewhat at odds with other
Indigenous populations.

Eschbach's survey of twentieth century patterns of American Indian
redistribution in the United States (Chapter 4) highlights this variation.
While emphasizing the diminished role of reservation to city migration in
the experience of most Indians, he is nonetheless more cautious in pro-
nouncing behavioural convergence. This is because links persist between

cities and reservations, although inadequacies in data restrict the analysis of mobility behaviour for urban populations.

Moving north, Frideres, Kalbach and Kalbach underscore the significance of treaty making and the reservation system in explaining the spatial redistribution of Canada's Aboriginal peoples (Chapter 5). Here, the shifting nature of Canadian government policy towards Indigenous populations is seen as paramount in controlling and directing population movement, especially between on- and off-reserve locations. In this regard, increased Indigenous self-determination places new emphasis on home-land territories as a residential base with potential impacts on mobility.

Part II (Data Issues and Analysis) narrows the focus to consider a crucial prerequisite for the analysis of population mobility: issues of data quality and application. As Indigenous populations are self-identified, for them to exist at all in a statistical sense requires two essentials. First, administrative mechanisms must be in place to ascribe and record Indigenous status. Second, there must be a willingness on the part of Indigenous people to be counted as such in order to differentiate them from the broader society. The degree to which these prerequisites combine to enable the compilation of demographic data varies greatly, both historically and between nations. Certainly, on a global scale, the statistical basis for a consistent description of Indigenous demography is tenuous at best.

Partly because they reside within developed nations, demographic statistics for the Indigenous populations under consideration here are com-paratively well developed. With varying degrees of coverage, and subject to changing interpretations of race and ethnicity, Indigenous people have been recorded in national censuses since 1881 in New Zealand, since 1870 and 1871 in the United States and Canada, and since 1901 in Australia. However, data issues remain a particular concern for the analysis of Indigenous mobility and in Chapter 6 Newbold outlines the salient issues across the different jurisdictions. A key problem is variability in the size, composition and distribution of populations due to changes over time in self-identification in census data. Allied to this are unique difficulties associated with conceptualizing and measuring changes in usual place of residence. This is compounded among Indigenous people by the prev-alence of circulation and multi-locale residence. Notwithstanding these difficulties, the team from the Strategic Research and Analysis Directorate of the Department of Indian and Northern Affairs in Canada (Norris *et al.*) find sufficient integrity and utility in census data to provide a detailed examination of recent flows of registered Indians between different segments of the Canadian settlement hierarchy. They also demonstrate how census data can be employed to quantify selectivity in the propensity to move (Chapter 7). Of particular interest is their application of census analysis to practical issues of policy relevance.

The chapters in Part III (Local Contingency) adopt more thematic, or case study approaches, to examine particular aspects of the interaction between Indigenous cultures and institutional frameworks that influence mobility behaviour. The papers in this part of the book engage a variety of disciplinary styles, almost ethnographic at times, to examine aspects of the intersection between mobility, cultural practice and structural forces at the local scale. In Chapter 8, Barcham, a Māori political scientist, employs the experience of his own Iwi (tribe) to illustrate the interdependence between urbanization processes and Māori political, social and organizational change, the most recent manifestation of which is return migration to tribal areas. In Chapter 9, sociologist Snipp, whose tribal background is Oklahoma Cherokee and Choctaw, explores the consequences of the American Indian diaspora for public policy. The special legal and political status of American Indians complicates jurisdictional responsibility for service delivery in the context of frequent mobility between reservations and cities. Added to this are the challenges facing tribal governments in maintaining culture, and authority structures among mobile groups.

Chapter 10 was written by demographer Alan Gray – one of his final manuscripts before untimely death in 2001. In this he details the historic processes of government control over Aboriginal mobility in Australia, involving authoritarian institutional concentration and dispersal. With the breakdown of direct controls over movement and subsequent urbanization, residential mobility is still highly regulated via the administration of public rental housing programmes. The outcomes are examined using a case study of several towns with significant Aboriginal populations in New South Wales.

The final two contributions are by anthropologists. In Chapter 11, Peterson debunks the myth of 'walkabout' as an aimless activity by charting the comings and goings of Aboriginal people across large swathes of the Australian continent with reference to kinship networks and ceremonial journeys. Mobility is conducted in Aboriginal domains for specific Aboriginal purposes. This theme is developed further, by Smith in Chapter 12, although he does this by exploring the intersection of Aboriginal life-worlds and state institutions in the context of emerging dispersed settlement on Aboriginal lands. His observation that the conditions of Indigenous articulation with the modern welfare state enable the continuance of social and spatial separation goes to the heart of a case for an Indigenous people's mobility transition. The degree to which this proves to be the case, or not, across the various jurisdictions provides a unifying thread throughout the book.

Despite drawing from a range of disciplinary sources, and, of course, from a diversity of regions and nation-states, common themes and approaches are evident. In particular, a tension is evident between the strength afforded to government and developmental agency in directing or enabling migration flows, on the one hand, and the primacy and continuity of Indigenous culture in giving expression to mobility outcomes, on the other. In effect, the patterns and trends in population movement observed provide a forceful manifestation of changing race relations in New World settings. Whether or not this also provides for a variant from demographic convergence is a theme explored by Bell and Taylor in Chapter 13 in their summary assessment of research themes that emerge from the individual papers. Although high mobility is a recognized stereotype of social and economic life among Indigenous peoples, understanding of the dynamics of movement remains fragmentary and appreciation of the national parameters of movement is only just beginning to emerge. It is clear, however, that distinct demographic processes have been, and are underway, and that these operate in response to unique structural and cultural imperatives. Furthermore, this is a characteristic clearly shared by the encapsulated societies of Australasia and North America, and the need for an improved understanding of these processes is heightened by the growing policy interest in Indigenous issues at both national and international levels.

References

Bell, M., Blake, M., Boyle, P., Duke-Williams, O., Hugo, G., Rees, P. and Stillwell, J. (2002) 'Cross-national comparison of internal migration: issues and measures', *Journal of the Royal Statistical Society A*, 165(3): 435–64.

Bodley, J.H. (1990) *Victims of Progress* (3rd edition), Mountain View: Mayfield Publishing Company.

Bongaarts, J. (2002) *The End of Fertility Transition in the Developing World*, Working Papers of the Population Council 161. New York: Policy Research Division.

Caldwell, J.C. and Caldwell, P. (2001) 'Regional paths to fertility transition', *Journal of Population Research*, 18(2): 91–118.

Caselli, G., Mesle, F. and Vallin, J. (2002) 'Epidemiologic transition theory exceptions', *Genus*, LVIII(1): 9–51.

Eschbach, K. (1993) 'Changing identification among American Indians and Alaska Natives,' *Demography*, 30(4): 635–52.

Fienup-Riordan, A. (1983) *The Nelson Island Eskimo*, Anchorage: Alaska Pacific University Press.

Frideres, J.S. (1983) *Native People in Canada: Contemporary Conflicts* (2nd edition), Scarborough: Prentice Hall.

Gray, A. (1985) 'Some myths in the demography of Aboriginal Australians', *Journal of the Australian Population Association*, 2(2): 136–49.

Hamilton, L.C. and Seyfrit, C.L. (1994) 'Coming out of the country: community size and gender balance among Alaskan Natives,' *Arctic Anthropology*, 31(1): 16–25.

Hogg, R.S. (1992) 'Indigenous mortality: placing Australian Aboriginal mortality within a broader context', *Social Science and Medicine*, 35(3): 335–46.

Hugo, G. (1988) 'Population transitions in Australia', in R.L. Heathcote and J.A. Mabbutt (eds) *Land Water and People: Geographical Essays in Australian Resource Management*, Sydney: Allen and Unwin, 162–212.

Jones, G. (1993) 'Is demographic uniformity inevitable?', *Journal of the Australian Population Association*, 10(1): 1–16.

Jorgensen, J.G. (1990) *Oil Age Eskimos*, Berkeley: University of California Press.

Kenen, R. (1987) 'Health status: Australian Aborigines and Native Americans – a comparsion', *Australian Aboriginal Studies*, 1987/1: 34–46.

Kunitz, S.J. (1990) 'Public policy and mortality among indigenous populations of Northern America and Australasia', *Population and Development Review*, 16(4): 647–72.

Kunitz, S.J. (1994) *Disease and Social Diversity: The European Impact on the Health of Non-Europeans*, New York: Oxford University Press.

Norris, M.J. (1990) 'The demography of Aboriginal people in Canada', in S. Shiva, F. Trovato and L. Driedger (eds) *Ethnic Demography: Canadian Immigrant Racial and Cultural Variations*, Ottawa: Carelton University Press.

Pool, I. (1986) 'The demography of indigenous minority populations', *Proceedings of General Conference, International Union for the Scientific Study of Population, Florence, 1985*, 4 vols, Liege: IUSSP, 1: 135–41.

Pool, I. (1991) *Te Iwi Maori*, Auckland: Auckland University Press.

Poulsen, M.F. Rowland, D.T. and Johnston, R.J. (1975) 'Patterns of Maori migration in New Zealand', in L.A. Kosinski and R.M. Prothero (eds) *People on the Move*, London: Methuen.

Robinson, V. (1992) 'The internal migration of Britain's ethnic population', in T. Champion and T. Fielding (eds) *Migration Processes and Patterns Volume 1: Research Progress and Prospects*, London: Bellhaven Press.

Ross, K. and Taylor, J. (2002) 'Improving life expectancy and health status: a comparison of Indigenous Australians and New Zealand Maori', in G. Carmichael with A. Dharmalingham (eds) *Populations of New Zealand and Australia at the Millennium*, Canberra: Australian Population Association.

Sahlins, M. (1999) 'What is anthropological enlightenment? Some lessons of the twentieth century', *Annual Review of Anthropology*, 28(1): i–xxiii.

Scott, K. and Kearns, R. (2000) 'Coming home: return migration by Maori to the Mangakahia Valley, Northland', *New Zealand Population Review*, 26(2): 21–44.

Skeldon, R. (1997) *Migration and Development: A Global Perspective*, London: Longman.

Snipp, M.C. (1989) *American Indians: The First of this Land*, New York: Russell Sage Foundation.

Taylor, J. (1991) 'Alaska and the Northern Territory: demographic change', in P. Jull, and S. Roberts (eds) *The Challenge of Northern Regions*, Darwin: North Australia Research Unit, Australian National University.

Taylor, J. (2003) 'Indigenous peoples', in P. Demeny and G. McNicoll (eds) *Encyclopedia of Population*, New York: Macmillan Reference.

Taylor, J. and Bell, M. (1996) 'Population mobility and indigenous people: the view from Australia', *International Journal of Population Geography*, 2(2): 153–70.

Wilson, C. (2001) 'On the scale of global demographic convergence', *Population and Development Review*, 27(1): 155–71.

Young, E.A. (1995) *Third World in the First: Development and Indigenous Peoples*, London: Routledge.

Zelinsky, W. (1971) 'The hypothesis of the mobility transition', *Geographical Review*, 61: 219–49.

Part I

International perspectives

2 Continuity and change in Indigenous Australian population mobility

John Taylor and Martin Bell

My attempt to supersede an undoubtedly strong legacy of traditionalism in the . . . literature concerned with Aborigines involves seeking to dissolve constricting dichotomies between persistence and change, between "traditionality" and (presumably) non- or post-traditionality. I assume neither to be more fundamental . . . , and begin where I assume dimensions of both persistence and change to have some relevance as subjective dimensions shaping social practice in the lives of Aboriginal people. I assume in my writing . . . the notion of an intercultural setting.

(Merlan 1998: 4)

This stance presents a challenge for mobility analysis, much of which is predicated on change. The very act of movement implies change in a place of residence, and when tied to indicators of social and economic change, as is so often the case in explanations of mobility, a self-reinforcing set of processes is invoked whereby mobility behaviour tends towards convergence at the national scale. The clearest illustration of this is seen in the observation that the most highly developed countries are also the most urbanized – hence the transfer of population from rural to urban areas is deemed to be an integral part of any process of development (Skeldon 1997: 2).

For Indigenous Australians, such unilinear argument is problematic since the setting for social action is truly intercultural, and decisions regarding mobility have been and are constantly shaped by a combination of persistence of the customary, and change due to external relations with the encapsulating state. Structurally, this places the socioeconomic position of Indigenous Australians at odds with conventional models of development. As Gray (1985: 143) points out, unlike populations of the Third World, Indigenous Australians live within a powerful polity and rich country, but in many respects at its fringes – they feel the exercise of power rather than its economic benefit. Enduring low socioeconomic status is the reality for most Indigenous Australians and this is as much a manifestation of inequitable power relations and marginalization in the midst of plenty as it is to do with any lack of development *per se*.

At the root of this discordance is the very stuff of changing race relations in Australia, with Gray (1985: 143) even suggesting that the only sustainable demography of Indigenous Australians is a political demography. In the past Indigenous people were directly impacted by the effects of interventionist government policy which provided for powers over their movement and place of residence. In contemporary Australia (post-1970), the social actions of Indigenous people are more appropriately viewed as a strategic engagement on their part wherein the principles and practice of self-determination enable persistence of the customary amidst pressures for change. This echoes Merlan's (1998) depiction of post-colonial relations as intercultural. Thus, rather than a dualism of development or tradition, one of the striking features of late capitalism at the outset of the twenty-first century is the survival of many Fourth World cultures by means of co-option and co-existence. In reflecting on this structural position, Sahlins (1999: x) has described the dominant mode as an "indigenization of modernity".

Accordingly, in this chapter, the importance of both Indigenous and state agency in directing mobility behaviour and outcomes is explored and emphasized. That change has occurred is undeniable, and we open by outlining transformations in the Indigenous population drawing attention to three elements of a proposed Indigenous mobility transition. These elements provide the framework around which the rest of the paper is constructed, the aim being to refine and modify each of these in light of growing evidence of continuities in mobility behaviour.[1]

Indigenous transformations

The Indigenous peoples of Australia comprise the descendants of disparate groups who occupied the continent for millennia prior to first European settlement in 1788. Those collectively referred to as Aboriginal peoples were widespread across the continent at this time, while Torres Strait Islanders of Melanesian origin were restricted to the archipelago across the strait of the same name between the Australian mainland and Papua New Guinea.

With the rapid development of Australia as a British colony, these Indigenous peoples became subjugated in phases by the inland advance of an emergent agricultural/industrial society and associated state apparatus – a process which by one interpretation was mostly complete by the end of the nineteenth century (Rowley 1970). Such was the vastness of the settlement frontier, however, that a group of Aboriginal people in the Great Sandy Desert of Western Australia did not have contact with the wider Australian society until as late as 1984 (Peterson 1986: 105).

For much of this colonial encounter, Indigenous Australians were expected by the state to disappear as a distinct people. As recently as the 1930s, this was expected to occur as a consequence of inexorable and excess mortality. In subsequent decades, with population decline in check, the means was social reclassification as Australian governments pursued a deliberate policy of cultural assimilation into mainstream society up until the 1970s. That the Indigenous population did not expire is underlined by very high rates of growth observed over the final three decades of the twentieth century due to natural increase and rising self-identification. Also over this period, the rights of Indigenous people to coexistence and a separate identity were legally enshrined, and official policy has moved away from enforced assimilation towards the granting of self-determination. Both of these developments accompanied a shift from the exclusion to the inclusion of Indigenous peoples in the provisions of the Australian welfare state (Altman and Sanders 1995).

One manifestation of such shifting colonial relations has been the formation of an "Indigenous" demography, distinct in many ways from the mainstream demographic profile, including that of mobility. Thus, while urbanization, and increasingly metropolitanization,[2] have accompanied the processes of economic development and globalization for Australians generally, the residential focus for Indigenous people remains away from major cities reflecting sociohistorical processes of Indigenous integration (or relative lack thereof) with the mainstream, mediated through a desire and capacity to sustain difference. In 1996, 63 per cent of the Australian population lived in metropolitan areas, compared to only 30 per cent of Indigenous people.

This is not to deny significant transformations in the Indigenous population. For example, as far as can be established from available data, a series of systematic fluctuations in fertility and mortality levels appear to have occurred, uneven over space and time, but ultimately comprehensive and uniform in effect. In broad terms, these describe an "Indigenous" demographic transition along the lines of the classical model observed generally for the Australian population. However, a significant revision is noted in the recognition of an initial, or pre-transition phase, of substantial depopulation contingent on first contact with non-indigenes (Smith 1980a).

Given the evidence of human habitation in Australia for tens of thousands of years, it is reasonable to assume that the population levels first encountered by European settlers were the product of long-term balance between birth and death rates. While most analysts suggest that this stationary state was due to sustained high birth and death rates, controversy surrounds the exact estimation of population size in 1788, with 300,000 as the preferred minimum figure, and 1 million as an upper bound (Gray and Smith 1983; Smith 1980a: 6).

Whatever the case, the negative impacts of an expanding colonial frontier led to a drastic decline in numbers due to reduced fertility and rising mortality. This decline was most rapid until 1890 with the population apparently stationary through to the 1930s at roughly 25 per cent of its original size, using the minimum estimate of the latter (Smith 1980a). The first sign of further transition appeared with a rise in the birth rate in the

1940s accompanied by a sudden and substantial drop in the mortality rate. This heralded a period of rapid population increase through to the 1970s. Though the current phase of transition is to a regime of lower natural increase, an increased propensity of individuals to self-identify in census counts as Indigenous Australian has maintained high overall growth rates (Gray 1997). At the turn of the millennium, the Indigenous population was estimated at 418,000 (Australian Bureau of Statistics 1998).

Alongside this demographic transition, the prospect of a parallel Indigenous mobility transition has been raised (Hugo 1988: 201–2), and part of the case for this lies in divergence from the mainstream transition. Long-term shifts in the mobility behaviour of the Australian population as a whole are reasonably well understood and documented. What is referred to as the "mobility transition" stems from the empirical observation of patterned and dominant regularities in the growth of personal mobility through space and time. This is seen as interacting with other elements of the demographic transition and to be synchronized with phases of economic and social development. For example, over the past 200 years in the shift from an economy concerned with primary production to one focused on tertiary and quaternary activities, the Australian population as a whole is seen as having moved from a period when rural migration to colonization frontiers was dominant to a period of net rural to urban migration, culminating in the present situation of counterurbanization and slight deconcentration of population combined with increased movement within and between cities. It is also apparent that migration patterns are dominated by the relations between State capitals and their hinterlands with relatively little net movement between States. At the same time, inter-State flows show distinct biases. Currently this also involves a northward and westward population drift from the older urban centres of southeastern Australia. which is partly production- and partly consumption-led (Bell and Hugo 2000).

In a suggested variant to this mainstream Australian mobility transition, Hugo (1988) has identified three overlapping phases of evolution in the patterns of Indigenous mobility and distribution since the arrival of Europeans. The first of these refers to the continuation of nomadic (or at least circulatory) movement associated with the maintenance of a customary hunting and gathering mode of production in those parts of the continent not expropriated by Europeans for cultivation or intensive grazing. The second phase identified was one of spatial concentration, whereby a nomadic population was encouraged and compelled to settle on reserves and mission stations up to the 1960s. The third phase refers to rapid postwar urbanization, especially since the 1960s.

While supporting this broad periodization into historic phases of mobility, this chapter seeks to highlight the importance of continuity and change in understanding patterns of Indigenous Australian population mobility over time and space. In so doing, modifications to the Hugo model are proposed.

First, while the persistence of circular mobility reflects continuity in customary land use practices for some Indigenous people in areas far removed from close settlement, the influence of customary practice is also evident in forms of circular movement that are more widespread and associated with the maintenance of social networks. At the same time, the changing demands of participation in mainstream institutions of work, education and training, as well as growing utilization of services, are seen as stimulating new patterns of circular movement.

Second, while the political economy of settlement distribution still serves to concentrate population in former mission and government towns, a significant countertrend has also emerged. This involves increased population dispersion to areas removed from existing settlements as family groups in many remote parts of the continent have gained legal access to their traditional lands and seek to resettle them.

Thirdly, despite the unequivocal rise in the enumerated urban population, some doubt is cast over the strength of rural–urban migration as a contributor to post-war urbanization. Reasons for this include evidence of increased self-identification of Indigenous peoples in census enumeration as the primary cause of growth in urban numbers, an overall net migration balance between metropolitan and non-metropolitan areas observed since at least the mid-1970s, and some ambiguity in the definition of urbanized populations due to high turnover.

Circular mobility

A recurring theme in the Australian literature on Indigenous population mobility is the recognition of circuits of population movement between places which combine to form functional regions. Networks of movement are reported from different parts of the continent and typically reflect localized linkages between sets of localities in both urban and rural settings (Altman 1987; Beckett 1965; Birdsall 1988; Bryant 1982; Sansom 1980; Taylor and Arthur 1993; Young and Doohan 1989). These are influenced by a mix of considerations that reflect persistence of the customary alongside change. Thus, mobility is shaped by continuity in land use practices, as well as in the importance attached to maintaining kin relationships. In some areas, such as Arnhem Land and central Australia, high levels of circulation are also associated with the on-going conduct of religious ceremonies (Altman 1987; Young and Doohan 1989; see also Chapter 11). At the same time, engagement with mainstream institutions is pervasive and, like the rest of the population, Indigenous people move to participate in education, training and employment, as well as to access essential services. It is often the balancing of these forces for continuity and change that lead many Indigenous people to engage in what one analyst has referred to as "multi-locale relationships" engendering high circular mobility (Uzzell 1976).

While the existence of such relationships is acknowledged, few attempts have been made to quantify the flows of population associated with them. This is partly due to the social porosity of communities and the difficulty of defining usual residential bases, but it also reflects the inadequacy of techniques aimed at capturing the dynamism of circular mobility (Taylor 1986). Certainly, low census-derived rates of mobility in remote regions of Australia conflict with the evidence from ethnographic studies which reveal high levels of circular movement (Altman 1987; Taylor and Bell 1996). This highlights common methodological issues concerning the limited value of fixed-period measures of migration and underlines the need to consider alternative ways of calibrating movement for populations where circulation is the predominant form of mobility.

Notwithstanding such data problems, the 1996 census still revealed that almost one-third of Indigenous people (29 per cent) changed their residence over the 12-month period prior to the census, a proportion which was much higher than the 18 per cent recorded for the rest of the population (Taylor and Bell 1999). While this higher rate of Indigenous mobility was partly due to the younger age profile of the Indigenous population, standardizing for this effect against the age distribution of the non-Indigenous population only reduces the Indigenous rate to 26 per cent. Thus, Indigenous people are almost 50 per cent more likely than the rest of the population to change residence over a one-year period. The fact that the Indigenous/non-Indigenous mobility differential is higher for the one-year period than for the five-year period (52 per cent compared with 43 per cent) also points to a greater propensity for Indigenous people to engage in repeat migration.

Circulation and land use

Evidence of circular mobility associated with contemporary hunting and gathering practices is available from case studies conducted in the north and centre of the continent (Altman 1987; Cane and Stanley 1985; Palmer and Brady 1991; Povinelli 1993; Young and Doohan 1989). The most detailed information is provided by Altman's (1987) description of a typical annual round of population movement among an eastern Gunwinggu band[3] at Momega outstation in the tropical savanna of north-central Arnhem Land.

Gunwinggu divide the seasonal cycle into six distinct periods involving three dry seasons and three wet seasons. Over a 12-month monitoring period between 1979 and 1980, each of these were observed to have an influence on population movement in the pursuit of subsistence harvesting of wildlife and land management. Thus, as in pre-contact times, seasonality remains a key determinant of residential locations and groupings, and the resultant spatial patterns of movement are shown in Figure 2.1. During the early and mid-wet seasons large areas of floodplain are inundated and the

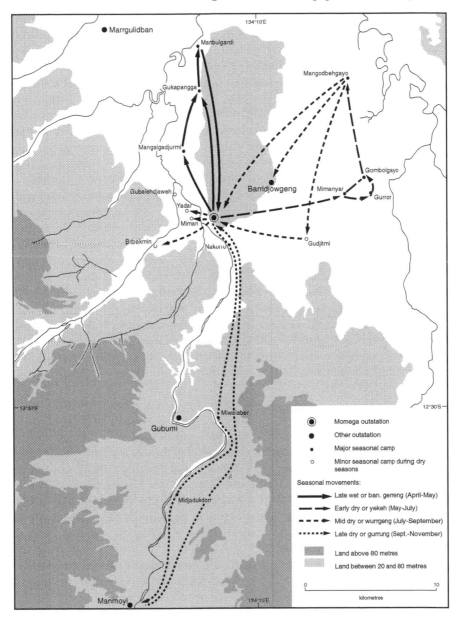

Figure 2.1 Circular mobility of the Momega band, north central Arnhem Land.
Source: adapted from Altman 1987: 24.

band population of some 40 people was concentrated at their main settlement of Momega on elevated ground. Mobility at this time was accordingly
low. As the first dry season encroached members of the band dispersed
until they began to regroup again several months later with the onset of the

first "early" wet season. The interim period was one of frequent mobility across an area of some 1,200 km² altering location according to resource variability. This circular pattern of movement is consistent with the tradi-tional range of Aboriginal bands in such tropical savanna regions that formed the basis of pre-contact spatial organization (Peterson 1986).

Of course, development in Arnhem Land, as elsewhere, has introduced townships, a market economy and transportation. As a consequence, new and more complex circuits of movement have emerged linking kinfolk over socially defined territories that are much larger than the subsistence range described above. Thus, the residential pattern of the Momega band includes not just the "immediate community" of outlying camps, but also a "wider community", comprising a section of the nearest service town and other distant outstations (Altman 1987: 102–7). In all, this settlement network circumscribes an area of some 5,600 km² within which the population of this one outstation interacts frequently with several other groups in a variety of distant localities. This merging of contemporary harvesting activity with the demands and opportunities of an introduced welfare economy has been described as a hybrid economy encompassing market, state, and customary sectors (Altman 2001).

One consequence of this mode of development, which is widespread across much of the tropical north and interior of the continent, has been a rise in the level and spatial extent of circular mobility with movement betw-een town and country enabled by vehicular and air access over large areas.

Circulation and social networks

Similar dynamics between social and settlement networks involving circular mobility over even wider areas are described for the Aboriginal populations of central Australia by Young (1981), Cane and Stanley (1985), Hamilton (1987), Young and Doohan (1989) and Peterson in Chapter 11. Such is the consistency of spatial interaction over large areas that Young (1990) refers to "mobility regions" as areas where common language, land custodianship, central place access, and kin location define the activity space of central Australian Aboriginal people.

Continuity of circular movement in the maintenance of social networks is also reported from a variety of urban settings. For example, from the early stages of urbanization Indigenous people sought to maintain links between their new residential bases and family in the hinterlands from which they were drawn. This was accomplished by engaging in frequent mobility between the two (Barwick 1962, 1964; Gale 1972: 90; Inglis 1964: 130–1). More recent studies in the Northern Territory indicate that circu-lation between urban areas and their rural hinterlands for the purposes of sustaining family ties remains an enduring characteristic of Indigenous mobility (Coulehan 1995; Drakakis-Smith 1980; Sansom 1980: 4–20; Taylor 1989; Young and Doohan 1989).

These circuits of movement have been variously described as "beats," "runs," and "lines" to reflect the pattern of mobility observed (Beckett 1965; Birdsall 1988; Sansom 1982: 122–30; Taylor and Arthur 1993), with the empirical construct described as "sets of places that constitute a social ambit for a person of no necessarily fixed residence but of a delimited countryside" (Sansom 1982: 125). So pervasive is this construct in ordering much of Indigenous mobility that the evidence from ethnographic analysis regarding regular patterns of movement is discernible at the aggregate level using census data.

Taylor and Bell (1996) have mapped the origin and destination of the largest (primary) and second largest (secondary) flows of Indigenous and non-Indigenous populations into and out of each of 54 regions nationwide using 1991 census data. On average, primary flows into each region accounted for 28 per cent of the total inflow, while the addition of secondary flows increased this proportion to 43 per cent. For both populations, the cumulative distribution of flows between regions was typically far from even, but skewness in the distribution was found to be much greater for the Indigenous population pointing to more intense localization of inter-regional movement. For example, the entire inflow of Indigenous movers to the Perth hinterland was accounted for by linkages with only 12 other regions. Even in Sydney, where Indigenous linkage with other regions was greatest, a large proportion of the total flow was still spatially restricted compared to that observed for the rest of the population.

Also of interest is the pattern of inter-regional flows. Unlike the majority population, inter-metropolitan primary flows were absent among Indigenous people, with regional networks much more in evidence and focused on the hinterlands of major cities. At the secondary level, the tendency for these regionalized networks of Indigenous flow was intensified, and more spatially widespread. Two basic patterns are discernible in Figure 2.2. The first involves circuits or "beats" of reciprocal movement. Some of these are localized – as in the Top End of the Northern Territory, the Perth and Adelaide hinterlands, as well as in many districts of New South Wales and Victoria. Others are more extensive – as in the desert regions of central and Western Australia. The second pattern is one of a series of "lines" or "runs" of movement most notable in the links between localities in central Australia, the far north of South Australia and on to Adelaide, as well as between regions along the Western Australian and Queensland coasts. The point to note here is that these spatial patterns are supported by evidence from the ethnographic record, as cited.

Circulation and services

One way of presenting a snapshot of the numbers involved in short-term migration, and the pattern of relocation that this creates, is by cross-tabulating census data on place of enumeration by place of usual residence.

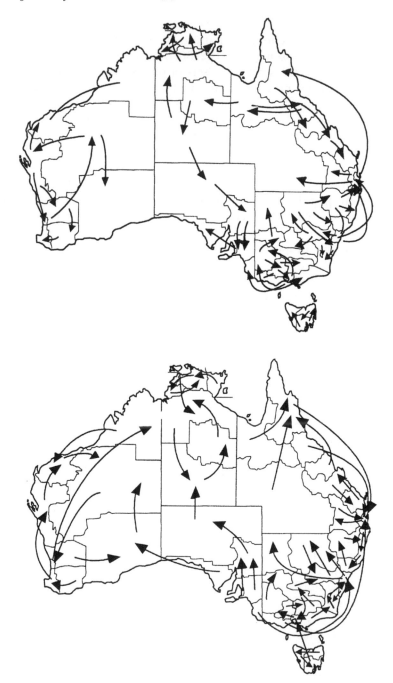

Figure 2.2 Secondary regional inflows and outflows of Indigenous movers, 1986–1991.

Source: Taylor and Bell 1996: 406.

In 1991, this revealed that 7 per cent of the Indigenous population was enumerated away from their usual place of residence, slightly higher than the 4.9 per cent recorded for the population as a whole (Taylor 1998).

Examination of this displacement across a 54-region nation-wide matrix revealed interesting patterns. While 57 per cent of Indigenous movers were involved in relatively local shifts within the same region, almost half (43 per cent) were temporarily absent in another region implying substantial long-distance displacement. Most of these latter transfers were between non-metropolitan and major city regions. Very little short-term Indigenous population transfer was recorded between major cities.

Of greater significance is the net effect of these movements in terms of temporarily adding to, or subtracting from, regional populations (Taylor 1998). Overall, net rates of inter-regional movement were low with gains or losses rarely exceeding 2 per cent of any region's usual resident population. This suggests that in most regions of Australia temporary transfers of population have only a minor demographic impact at any one time, even though a considerable number of individuals are likely to move during a given year. At the same time, a fairly regular pattern of regional net gains and losses is evident. It is noticeable, for example, that all major city regions experience net gains, while the majority of non-metropolitan regions (80 per cent) record net losses – a pattern of short-term movement opposite to that observed for the total population (Bell and Ward 2000). The indication is that at any one time a proportion of the Indigenous population of major cities and regional centres around the country comprises individuals from non-metropolitan areas who are only short-term residents.

In 1991, the degree to which this was so varied considerably from just over 2 per cent in Sydney to over 11 per cent in Darwin. In each case this net shift of short-term movers reflects the spatial concentration of higher-order services in relation to the distribution of much of the Indigenous population. Thus, for many Indigenous people, access to banks, hospitals, prisons, government offices, employment opportunities, public housing, education and training institutions involves considerable travel and time away from home.

The effect of this mobility to service centres is to create a pool or catchment of population around each service town. Some sense of the size of these population catchments, and their spatial extent, is provided by data from the 1999 Indigenous Community Housing and Infrastructure Needs Survey (CHINS) conducted by the Australian Bureau of Statistics (ABS). This recorded the nearest town that members of each community in remote Australia travelled to in order to access services such as banking and major shopping. A total of 96 service centres across remote Australia were identified servicing 1,100 smaller communities with a collective population of 80,000. Figure 2.3 provides an indication of the spatial pattern of their catchment areas.

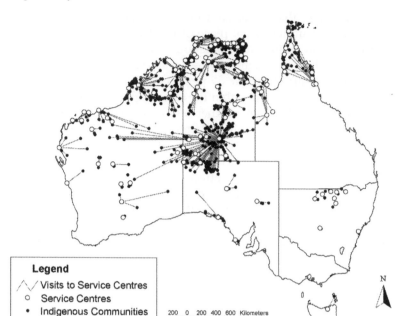

Figure 2.3 Journey to service centres: Indigenous communities in remote Australia, 1999.

Source: Taylor 2002.

The map highlights the major role played by Alice Springs in servicing vast areas of central Australia. In all, Alice Springs (population 24,000) services some 260 small Indigenous communities encompassing a combined estimated population of 15,000. Moving north, Katherine and Darwin emerge as other major regional centres of attraction, while Cairns stands out in north Queensland. In Western Australia, a string of smaller catchment areas are evident. In each case, the primary direction of circular movement for service access is illustrated and it is significant that not all populations access their nearest service centre. This is partly a function of variable transportation links, but in some instances it reflects patterns of cultural affiliation, as among desert peoples in the centre of Western Australia who prefer to travel long distances eastwards to towns in the Northern Territory. Also of note are the vast distances traversed within many of the catchment areas involving round trips of 1,000 km or more.

Movement of individuals to access services can also generate ancillary flows. For example, in many parts of remote Australia health care strategies are predicated on the air evacuation of Indigenous patients to urban centres for hospital and other health services. Research among the Yolngu of northeast Arnhem Land suggests that this transfer of medical patients represents only a fraction of total health-related mobility as it also generates much larger, related movements, primarily of kinfolk anxious to fulfil cultural obligations to "keep company" with their sick relatives for the duration of treatment (Coulehan 1995). Such transience is enhanced by the fact that the threshold for hospital admissions is often lowered in an attempt to improve morbidity and mortality rates among remote populations, as well as to compensate for the lower effective health service delivery in Aboriginal communities.

This spatial dichotomy between the concentration of services and dispersion of population over vast distances raises a number of questions regarding access and equity (Taylor 1998). For example, if the residence pattern of many Indigenous people is multi-local, where are services optimally located? Should services be replicated to cater for frequent movement between places? If urban areas are net recipients of temporary sojourners, should urban services be augmented to compensate for additional loads, or should attempts be made to decentralize service delivery to overcome the need for movement?

Circulation and socioeconomic status

A substantial and growing literature is available detailing the relatively low socioeconomic status of Indigenous Australians and examining underlying causes (Altman 2000; Altman and Nieuwenhuysen 1979). Viewed sequentially, these analyses reveal intractability in the economic plight of the Indigenous population. A common thread relates to underlying determinants that remain focused on locational disadvantage, poor human capital endowments and the historic legacy of exclusion from the mainstream provisions of the Australian state.

However, the extent to which frequent and temporary mobility might be causally related to low socioeconomic status is poorly understood. On the one hand, there are indications that the persistent mobility of many Indigenous people may influence the level and nature of their interaction with mainstream institutions, including those that impact on socioeconomic status. For example, frequent mobility has been found to contribute to difficulties in ensuring delivery of social security benefits (Sanders 1999: 93–4), and to reduced rates of school attendance (Northern Territory Department of Education 1999: 146–8). On the other hand, this very lack of engagement with mainstream institutions, in part occasioned by low socioeconomic status, may itself contribute to higher mobility by freeing individuals to pursue other priorities of more traditional concern.

In effect, social and economic factors operate as a dynamic complex of attractions and pressures in different locations to induce a cycle of movement between them. This situation, reminiscent of Merlan's (1998) intercultural space, is conceptualized in Figure 2.4 which highlights a consensus of migrant motivations from a survey of rural and urban households in the Northern Territory and conveys something of the mix of considerations which govern local Aboriginal mobility (Taylor 1989). While the motivational categories shown are not exhaustive, nor are they mutually exclusive; they are selected to emphasize the intercultural nature of the factors influencing mobility decision-making.

Thus, individuals in remote Aboriginal settlements are subject to a range of pressures which may serve to induce their movement to an urban area. Foremost among these is the physical separation of kinfolk between town and country and the disproportionate access to social services experienced by rural dwellers. In addition, most, if not all, large Aboriginal communities in Australia are artificial concentrations of population established primarily as a means of controlling Aboriginal people under policies of protection and assimilation. As such, they required no modern economic base, nor, for the most part, have they subsequently acquired one, at least not in a manner that is sustainable beyond the provisions of the welfare state. Thus, movement to urban centres offers the only prospect of accessing mainstream employment. Once in town, however, several factors tend to work against long-term residence. Aside from the fact that this may not have been sought anyway, a lack of adequate and appropriate accommodation, social problems created by access to alcohol, and the tensions of living in a multicultural, largely non-Aboriginal town, all serve to reinforce the attractions of remote settlements. As for employment prospects, these are generally undermined by characteristically low educational status and skill levels.

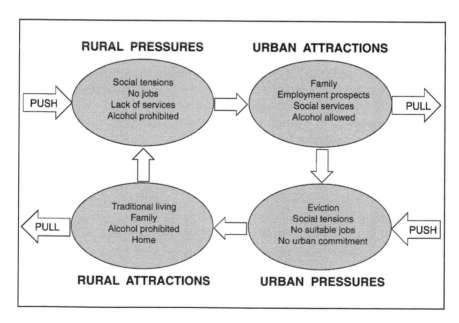

Figure 2.4 Push–pull factors in Indigenous rural–urban circular mobility.
Source: Taylor 1989: 51.

The degree of relative detachment from mainstream institutions portrayed in Figure 2.4, and the inducement to frequent mobility that it represents, appears to be reflected in the profile of Indigenous mobility by age and sex. As shown in Figures 2.5 and 2.6, the shape of the age profile of Indigenous mobility is similar to that observed for all other Australians with movement rates peaking in the 20–29 years age range followed by a sharp decline, but with a slight rise in the retirement ages. A secondary peak is also evident among infants and children, reflecting the migration of family groups. Despite these similarities, a significant point of difference evident for one-year rates is the much higher mobility of Indigenous males and females compared to their non-Indigenous counterparts (Figure 2.5). This is especially so among children and youth, and those in middle and old age. Thus, around 34 per cent of Indigenous infants change their usual place of residence each year, compared to only 23 per cent of non-Indigenous infants. Likewise, around 25 per cent of Indigenous children of compulsory school age change their usual place of residence each year compared to only 15 per cent of non-Indigenous school-age children. In the years of peak movement, between 20 and 29 years, movement rates are much closer ranging between 35 and 45 per cent. At older ages mobility rates fall away for both populations, but the differential between them increases such that Indigenous rates are around 10 to 20 percentage points above those for non-Indigenous people. These persistently higher Indigenous mobility rates at young and older ages are

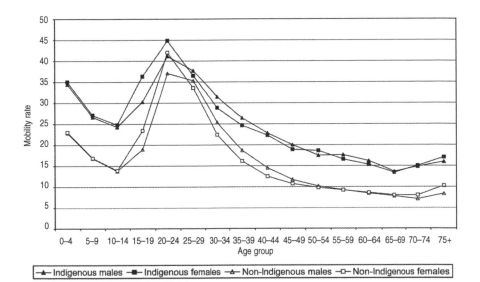

Figure 2.5 Age and sex profile of Indigenous and non-Indigenous mobility rates, 1995–1996.

Source: 1996 ABS Census of Population and Housing, unpublished data.

interesting as they suggest that the usual life-cycle events of schooling, mortgage repayment, and career development that generally serve to dampen mobility, have a much weaker influence for Indigenous people.

While five-year mobility rates display the same life-cycle variations (Figure 2.6), a key point of difference is the much flatter age profile for both Indigenous males and females. For the population in general, the peak in the age profile of migration has been firmly linked to the combined influence of departure from the parental home, the start of tertiary education, entry into the labour force and the establishment of independent living arrangements (Bell 1992: 92). Thus, the much flatter profile of Indigenous mobility may reflect the much lower labour force participation observed for Indigenous people at ages when job search and job mobility are primary factors in population movement for the rest of the population. It may also be a measure of the impact of the federal government's Community Development Employment Projects (CDEP) – a work-for-the-dole scheme for unemployed Indigenous people which requires a weekly commitment to some employment activity for as much as one-fifth of the Indigenous workforce. Also, the establishment of independent living arrangements is likely to be less of a stimulus for migration among young Indigenous adults in a cultural setting that places emphasis on maintaining extended kinship ties. Shortage of available housing to accommodate new household formation, combined with high dependence on rental housing, is an additional constraint (see Chapter 10).

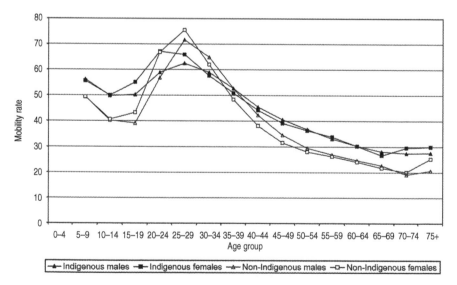

Figure 2.6 Age and sex profile of Indigenous and non-Indigenous mobility rates, 1991–1996.

Source: 1996 ABS Census of Population and Housing, unpublished data.

Population concentration and dispersal

With gathering pace from the late eighteenth through to the early twentieth centuries, the original distribution of small groups of Indigenous peoples across Australia largely disintegrated in the face of expanding colonial frontiers (Reynolds 1987, 1992; Rowley 1970). In the areas most desired for European settlement, expansion of the frontier brought disease, starvation, neglect and genocide for the Indigenous occupants. Surviving populations were mostly displaced, and under the direction of protection policies formerly dispersed groups were relocated into concentrated pockets on mission stations, government settlements and reserves, often far removed from ancestral lands. Thus, people accustomed to living in small groups with freedom to move became confined in large sedentary groups, and their experience of settlement and urban life began.

Much of the history of this early centralization has emphasized the effect of coercion over spontaneous integration. While opportunistic movement to emerging urban centres did occur, the most significant redistribution resulted as a direct consequence of appropriation of land and of government policy and practice. Of particular effect were the provisions under ordinances of protection and welfare which, from the late nineteenth century until as late as the 1960s, declared Indigenous people as wards of the state and provided powers to control their movement and place of residence leading to the consolidation of highly institutionalized settlement. Regional histories of government-enforced migration are available (Anderson 1986; Barwick 1964; Beckett 1965; Brady 1999; Gale 1967, 1972; Inglis 1964; Long 1970), while Rowley's (1970, 1971a, 1971b) comprehensive analyses of Aboriginal policy and practice provide the essential background nationwide. However, within this broad coercive framework, evidence also points to Aboriginal people as active and willing participants in gradual migration to towns and settlements (Brady 1987; Long 1989; Read and Japaljarri 1978). It is also suggested that the apparent "coming in" to settlements was also a case of European settlement "going out" to places that were already populated by Aboriginal people on a seasonal basis (Baker 1990: 59).

Whatever the case, incorporation into wider economic structures clearly generated its own migration dynamics. One factor, in particular, which sustained a long-standing rural bias in the distribution of Indigenous people across the country was their heavy dependence on the agricultural sector for employment. From the 1960s onwards, this largely seasonal employment as stockmen, fruit pickers and general agricultural labourers was steadily eroded due to structural change in the industry and mechanization rapidly displacing Indigenous workers. Castle and Hagan (1984) and Lea (1987: 66–7; 1989: 65–80) have identified increased migration from rural areas as one consequence of this change.

Government welfare policies also contributed to rural–urban migration. For example, resettlement schemes initiated in New South Wales in the

1960s had the aim of encouraging families from the most depressed rural areas to migrate to urban centres (Mitchell and Cawte 1977; see also Chapter 10). The major reasons for relocating given by migrants were the availability of better housing and employment opportunities. Other pressures included the discriminatory use of unemployment benefits in favour of those willing to migrate (Ball 1985: 5). A lack of rural employment opportunities has also been stressed in the case of Torres Strait Islanders who, by virtue of their original location in the remote archipelago of the Torres Strait, have undergone a more visible redistribution, notably to urban centres on the Queensland coast, in their search for jobs (Taylor and Arthur 1993).

Return to country

While the spatial concentration of Indigenous peoples is a lasting legacy of colonization, in the final decades of the twentieth century there was (and continues to be) a revival of settlement in areas formerly depopulated, especially in the remotest parts of the continent. Reference to remoteness and "the outback" is long-standing in Australian regional analysis. This draws attention to a distinction in social and economic geography between closely settled areas and the 80 per cent of the continent that is sparsely settled, with economic development and service provision severely impeded in the latter by force of relative locational disadvantage, low accessibility, and a specialization of economic activity (Holmes 1988). Along similar lines, Rowley (1971b) has drawn distinction between "colonial" and "settled" Australia in recognition of the much higher proportions of Indigenous people in remote areas, and the somewhat different manner of their incorporation into wider social and economic structures. Indeed, away from the larger mining towns and service centres of the outback, it is possible to talk of Indigenous domains in the sense that the Indigenous population by and large constitutes the public.

For some decades now, demographic trends in remote Australia have been volatile. Since 1981, the Indigenous share of the total population within an area approximating the Australian outback rose steadily from 12 per cent to almost 20 per cent in 1996 (Taylor 2000). This occurred as a consequence of differential population dynamics – the Indigenous population is much younger in age profile, and has experienced a much higher rate of natural increase than the population in general. Also, many Indigenous people in remote areas reside close to their ancestral lands and their attachment to such places is reflected in a relative lack of net out-migration (Gray 1989; Taylor 1992; Taylor and Bell 1996, 1999). This contrasts with the historically more recent and ephemeral non-Indigenous settlement of the outback with the experience of recent decades being one of an ageing population and generalized out-migration leading to population decline in many non-metropolitan districts (Bell 1995; Bell and

Hugo 2000; McKenzie 1994). Since 1981, the Indigenous population in remote areas of Australia has grown by 23 per cent. By contrast, since 1986, overall non-Indigenous population growth has been negative (Taylor 2000).

The effect of this on the relative distribution of the Indigenous population is shown in Table 2.1 which indicates Indigenous and total population numbers according to regions constructed by the ABS on the basis of accessibility to population and services. The vast majority of Australians (87 per cent) live in major cities and their immediate hinterlands with high proximity to the widest range of services. At the other extreme, less than 3 per cent of Australians are located in remote and very remote areas where service access is severely restricted. By contrast, almost 30 per cent of the Indigenous population remains resident in remote and very remote areas where they account for a significant share of the total population, certainly when benchmarked against their 2.1 per cent share of the national population.

A key factor underpinning these quite different distributional patterns is a substantial return of land across remote Australia back to Aboriginal ownership since the 1970s, with the prospect of more to come via land purchase and native title claim. According to Pollack (2001), Indigenous landholdings in 1996 accounted for at least 15 per cent of the Australian land mass. The vast bulk of this area was found in remote Australia, mostly in the Northern Territory, followed by Western Australia, South Australia and Queensland. In the Northern Territory, these substantial holdings comprised half of the land area.

This restitution of land into traditional ownership is an important element of the post-productivist transition in Australia's rangelands since newly recognized land values often lie outside the market economy and are more culturally based (Holmes 2002). One such response is manifest in the emergence of a distinct settlement structure on Aboriginal lands involving the reconstitution of numerous, dispersed, small, and discrete[4] Indigenous

Table 2.1 Distribution of Indigenous and total populations by remoteness category, 1996

Remoteness category	Total population (millions)	Indigenous population (no.)	Total population (%)	Indigenous population (%)	Indigenous share of total (%)
Major City	12.1	113,825	66.1	29.5	0.9
Inner Regional	3.7	71,825	20.5	18.6	1.9
Outer Regional	1.9	91,844	10.7	23.8	4.7
Remote	0.30	36,760	1.8	9.5	11.2
Very Remote	0.17	71,602	0.9	18.6	41.6
Total	18.3	385,856	100.0	100.0	100.0

Source: 1996 ABS Census of Population and Housing, customized tables.

communities. This has occurred across much of the Northern Territory, Western Australia, the far north of South Australia, and north Queensland and has been referred to somewhat poignantly in one report as a "return to country" (Commonwealth of Australia 1987), although others prefer to downplay the "return" element placing emphasis more on an expansion of settlement options with continuing links to regional centres (see Chapter 12).

While a population presence has always been sustained in remote parts of the continent, this has been enhanced since the early 1970s by a combination of factors including: the granting of land rights; increased pressure on traditional lands from miners and other resource developers; direct access to government funds for vehicles, capital equipment and infrastructure; and the desire to re-establish living arrangements in smaller communities away from the many social problems associated with daily life in centralized rural townships. These have stimulated a purposeful shift of population toward smaller scale, dispersed settlements referred to as "outstations" or "homeland centres" (Altman and Nieuwenhuysen 1979: 76–100; Altman and Taylor 1989; Commonwealth of Australia 1987; Coombs 1974; Coombs *et al.* 1982; Taylor 1992).

Across the continent, a total of 1,290 discrete Indigenous communities were identified in 1999 with a total population of 110,000. The vast majority of these were outstation settlements of less than 50 persons – essentially family groups – although other larger clusters represent Indigenous living arrangements formerly constituted as government and mission settlements. They are located in all States and Territories but are found overwhelmingly in remote areas of Queensland, Western Australia, South Australia and the Northern Territory with major concentrations on Indigenous-owned lands in central Australia, the Kimberley region, the Top End of the Northern Territory and Cape York Peninsula.

From the bureaucratic perspective of those seeking to provide services and achieve social and economic equity goals, such a focus by Indigenous people on utilizing and residing on Aboriginal lands is often construed as a retrograde step on the grounds that it serves to reinforce the locational disadvantage of a group already severely disadvantaged according to standard social indicators. From an Indigenous cultural perspective, however, ongoing dispersion of population can be seen as representing the spatial optimum in a locational trade-off which is aimed at balancing a range of cultural, economic, social and political considerations. Such a trade-off involves reduced access to urban-based mainstream labour markets, opportunities for education, training and income generation as well as access to better housing and other social facilities. To the extent that these are perceived as losses, they are set against the not insignificant social, cultural and economic gains acquired from renewed residence on Aboriginal lands (Altman 1987, 2001).

Post-war urbanization: real or imagined?

Notwithstanding the retention of Indigenous populations in remote areas, one of the more obvious transformations in the second half of the twentieth century was a shift in the balance of continental geographic distribution away from remote and rural areas in favour of urban and metropolitan centres, and consequently from the north and west to the south and east of the country. Over the longer term, as noted, this redistribution may be viewed as an outcome of the incorporation of Indigenous people into the expanding and modernizing Australian state. Over the shorter term, since the 1960s, uncertainty exists as to whether it is spatial mobility, or rather ethnic mobility, that has been the main contributor to increased urban numbers.

In the 1960s, a series of survey- and census-based analyses highlighted what had been perceived as occurring for some time – that the Indigenous population resident in major cities had been growing rapidly due to net migration gains from small towns and rural areas. In the definitive study, based on Adelaide, it was argued that movement from mission and government reserves was stimulated by a search for employment opportunities and the attraction of better social services (Gale 1967, 1972; Gale and Wundersitz 1982). Once metropolitan links were established, movement out of rural areas was sustained by a process of chain migration involving kin networks. Similar sets of push and pull factors, with particular emphasis on the search for employment, were reported from migrant surveys in other cities (Beasley 1970; Brown *et al.* 1974: 19; Burnley and Routh 1985; Smith and Biddle 1975: 42–53).

The clearest example of such rural–urban migration occurred among Torres Strait Islanders (Taylor and Arthur 1993). Until the end of World War II, Torres Strait Islanders were restricted by law and administrative arrangements to residing in scattered communities across the Torres Strait archipelago. This is despite notable exceptions, such as those Islanders who seasonally ventured further afield across northern Australia as crews on pearling luggers. Due to subsequent out-migration for employment, however, and natural increase of the Islander population on the mainland, this pattern of distribution is now almost completely reversed. In 1996, almost 85 per cent of all Torres Strait Islanders were resident on the mainland. Furthermore, the pattern of settlement which has emerged from this redistribution is quite distinctive, being focused primarily on the larger urban centres of North Queensland, as well as metropolitan Brisbane and Sydney. Altogether, three-quarters of Torres Strait Islanders now reside in mainland urban centres.

On the face of it, the migration of Aboriginal and Torres Strait Islander people to towns and cities reported in the 1950s and 1960s, appears to have accelerated in subsequent decades. Thus, at the 1971 census (the first to include a self-identified Indigenous population in the overall count of the Australian population), 56 per cent of those who declared themselves of Aboriginal or Torres Strait Islander origin were resident in a rural area. At the 1996 census, this figure was reduced to only 27 per cent. Accordingly, the proportion of the Indigenous population resident in urban areas rose from just 44 per cent in 1971 to 73 per cent in 1996. At the same time, 30 per cent of Indigenous Australians are now resident in major cities, and while this remains considerably less than the total population (63 per cent), it nonetheless represents a doubling in proportion since 1971.

If anything, these figures understate both the extent and rapid rise in urban numbers, especially in terms of proximity to metropolitan centres and large cities. The criteria used to classify statistical units as urban or rural are based on measures of population density, land use and spatial contiguity. This means that many people who may reasonably be regarded as forming part of a city region are not classified as urban dwellers. By adding peri-urban areas to the calculation, 27 per cent of the Indigenous population was classified as resident in major city regions in 1991. By 1996, this figure had risen to 36 per cent.

A revisionist view

While there is no doubting the steady rise in urban numbers and the empirical evidence pointing to rural out-migration, a revisionist view of this redistribution has emerged. This regards post-war migration to major cities as only a temporary wave during the 1950s and 1960s and one which contributed less to Indigenous urban population growth than previously claimed (Gray 1989: 130–3). Suspicion that much of the apparent shift in population distribution since the 1960s could have been due to an increased tendency for city-based Indigenous people to self-identify in census enumerations, as much as to net migration, has been raised by Smith (1980a: 252; 1980b: 202) and Gray (1989: 130). This is based on the fact that early studies of Indigenous urbanization focused solely on flows towards cities with no corresponding statistics on counterstreams of people who may have been leaving metropolitan areas.

The point is addressed in detail by Gray (1989) who demonstrated emphatically that if migration were ever a major factor leading to an increased Indigenous presence in major cities then, from 1976 onwards, it was far less so. The same point has been acknowledged by Gale and Wundersitz (1982: 96) who noted that movement patterns based on Adelaide were not unidirectional, but included a good deal of movement back out to country areas. They also concluded that migration flows to the city peaked during the 1960s with subsequent growth in urban areas due more to the effects of natural increase (Gale and Wundersitz 1982: 39). For the contemporary period, Gray (1989) and Taylor and Bell (1996, 1999) have commented on the low overall effectiveness of migration flows between metropolitan and non-metropolitan areas, with the 1996 census indicating slight overall net gain to capital cities in contrast with the previous intercensal period which showed a slight net loss (Table 2.2).

However, this aggregate net gain of Indigenous population to capital cities was by no means uniform. Sydney, for example, experienced a sizeable net loss of Indigenous people to the rest of New South Wales as well as to other States. This also occurred in previous intercensal periods back to the 1970s. Melbourne has also recorded net losses of Indigenous population both to the rest of Victoria and to other States. In the other major cities of Brisbane, Adelaide, Perth and Canberra, net migration gains have been consistently recorded. Interestingly, this mirrors the pattern observed for the Australian population as a whole, though for quite different reasons.

Population turnover

In contrast with the pattern observed generally, high inter-regional turn-over rates for the Indigenous population, involving half or more of a region's population, tend to be associated with metropolitan areas, notably

Table 2.2 Indigenous intrastate and interstate migration rates (per thousand) for capital cities and rest of state, 1991–1996

Region	Intrastate migration			Interstate migration		
	In-	Out-	Net	In-	Out-	Net
Sydney	62.1	83.7	−21.6	34.4	48.1	−13.7
Rest New South Wales	43.0	32.9	10.1	33.8	54.2	−20.4
Melbourne	47.0	60.0	−13.0	67.6	80.0	−12.4
Rest Victoria	59.7	46.8	12.9	84.8	120.9	−36.1
Brisbane	120.6	94.0	26.6	90.3	40.8	49.5
Rest Queensland	26.7	34.2	−7.5	52.3	34.4	17.9
Adelaide	80.5	74.1	6.4	102.1	89.3	12.8
Rest South Australia	62.1	67.5	−5.4	65.2	63.3	1.9
Perth	134.6	95.6	39.0	53.8	36.7	17.1
Rest Western Australia	45.1	63.5	−18.4	25.7	28.4	−2.7
Hobart	100.9	78.3	22.6	38.1	60.1	−22.0
Rest Tasmania	38.8	50.0	−11.2	30.5	53.4	−22.9
Darwin	144.3	99.5	44.8	116.2	131.0	−14.8
Rest Northern Territory	17.4	25.2	−7.8	28.6	31.0	−2.4
Canberra	n.a.	n.a.	n.a.	313.9	228.7	85.2
Total capital cities	90.3	83.0	7.3	70.1	62.3	7.8
Total non-metropolitan	35.6	38.8	−3.2	40.8	44.3	−3.5

Source: 1996 ABS Census of Population and Housing (unpublished data).
Notes:
n.a.=not applicable.
In-, out- and net rates are derived from arrivals, departures and net movement for each region divided by the mean intercensal usual resident population in each region expressed as parts per thousand.

Brisbane, Perth, Melbourne and Adelaide (Gray 1989; Taylor and Bell 1999). The Indigenous populations of Canberra and Darwin also display relatively high rates of turnover, even in these typically migrant cities. For example, the rate of turnover of the Indigenous population of Canberra in 1996 was 1.5 times that of the rest of the population.

This high turnover is attributed largely to movement generated between cities and their hinterlands, as opposed to involving inter-metropolitan movement. Gray (1989: 133) has made the point that this tends to undermine the notion of an "urban Aboriginal population" as distinct from any other. He goes further to suggest that Aboriginal people in the city are not just similar to those in country areas – to a large extent they are the same people spatially displaced at different stages of their lives (Gray 1989: 133).

The basis for this assertion stems from Gray's analysis of the age-specific pattern of net flows in and out of cities with two overlapping patterns of urbanization observed. The first is evident in the large metropolitan centres of Sydney and Melbourne and involves a cycle of young single people moving to the city, then returning to the country maybe ten years later taking their new families with them. The second pattern is focused on the smaller cities of Adelaide and Perth and involves more permanent migration, possibly owing to the existence of more active Aboriginal housing programmes in those cities. In all States, net in-migration to cities is concentrated in the 15–24 age group, highlighting an economic imperative in the context of education, training and job search, while out-migration at older ages reflects difficulties in securing family housing. The common socioeconomic determinant here is the much greater reliance of Indigenous people on access to housing via the public sector (Drakakis-Smith 1981; see also Chapter 10).

Net migration

As might be expected, a strong positive relationship exists in Australia between regional net migration gain and regional population growth. Put simply, regions that experience growth in population do so largely because of net gains from migration. Conversely, those experiencing decline do so mostly because of net migration losses. While the form of this relationship also holds for the Indigenous population, the association is much weaker with many regions, especially those focused on metropolitan areas or with mostly urban populations, experiencing population growth (substantial at times) far above expectation given their net migration rate. This is underlined by the fact that some regions display high population growth despite experiencing negative net migration. For example, between 1991 and 1996, the metropolitan area of Sydney experienced a 31 per cent increase in its Indigenous population despite experiencing a net migration loss (Taylor and Bell 1999). Overall, this low association can be traced to non-demographic factors in population growth, mostly an increased propensity for individuals to identify as Indigenous in the census.

Volatility in census counts is almost a defining feature of the self-identified Indigenous population. Since 1971, intercensal change in recorded numbers has been negative on one occasion, and often excessively positive (Taylor 1997). Given the tendency for high error of census closure among self-identified populations (Passel 1996), successive census data effectively report the characteristics, including those pertaining to mobility, of different populations. While some scope exists for estimating the compositional impact of newcomers on the population using fixed population characteristics such as age left school (Eschbach *et al*. 1998), for characteristics that are variable over time, such as mobility status, this is simply not possible. The fundamental and unresolved difficulty for migration analysis, and particularly in tracking urbanization trends, therefore lies in estimating the extent to which growth in urban population reflects spatial, as opposed to ethnic, mobility.

Indigenous mobility transition

Although high mobility is a recognized characteristic of social and economic life among Indigenous Australians, understanding of the dynamics of movement remains incomplete, while the measurement of national parameters is only just beginning to emerge. It is clear, however, that distinct demographic processes have been and are underway, and that these operate in response to unique structural and cultural imperatives. As Borrie (1994: 6) suggests, it is appropriate to approach the demographic analysis of Indigenous Australians as a case study of the outcomes of "white" imperial settlement in "New World" countries. Implied in this observation is a notion, common to Indigenous peoples worldwide, of imposed change.

What also needs emphasis, certainly in the Australian context, is the fact that change is mediated through Indigenous agency. This is reflected in the continuities evident in mobility behaviour and outcomes. This chapter has sought to highlight the importance of understanding patterns of Indigenous Australian population mobility over time and space as an on-going tension between continuity and change. In so doing, three major modifications to an Australian model of Indigenous mobility transition have been presented. The first phase which refers to the continuation of circulatory movement associated with the maintenance of a customary hunting and gathering mode of production should be expanded spatially to include all circular movements associated with the maintenance of customary kinship. The second phase, one of spatial concentration whereby a nomadic population was encouraged and compelled to settle on reserves and mission stations up to the 1960s, needs to now incorporate a post-1960s phase of population dispersal in remote areas consequent on the removal of controls over movement and the granting of land rights. The third phase refers to rapid post-war urbanization, especially since the 1960s. This requires revision to take account of the effects of ethnic mobility in the urbanization process as well as indications of high urban population turnover.

The significance of these modifications stems from the lack of conformity they present to the notion of inexorable drift to urbanization as envisaged in the hypothesized Zelinsky sequence of mobility transition (Zelinsky 1971), and to Skeldon's (1997: 2) related view that the transfer of population from rural to urban areas is an integral part of any process of development. The relative lack of urbanization observed among Indigenous people in Australia, the contemporary fragmentation of their rural settlement, and the persistence of their short-term mobility over long-term migration, are all observed in an advanced phase of Australian development, yet all run counter to this proposition of inexorable urban drift. At one level, this reflects the lack of Indigenous integration with mainstream institutions, at another it demonstrates an ongoing capacity of Indigenous people to sustain difference. What remains unclear is the manner and extent to which these might be related.

Thus, as Skeldon (1997: 199) acknowledges, encapsulated cultures in developed countries encounter a quite different set of structural circumstances than, say, populations of Third World countries (Skeldon 1990): namely, that their Indigenous rights have been increasingly recognized and protected, and they have some means to pursue self-determined lifestyles away from urban and modernizing influence as underwritten by welfare provisions within liberal democratic states, or by force of their own economic base on traditional lands. As demonstrated, this interplay has forged unique outcomes both in the nature of their mobility and in the spatial patterning of migration paths which, together, lend strong support to the notion of an Indigenous mobility transition.

Notes

1 We are grateful for comments on early drafts of this paper from Jon Altman, Diane Smith, Will Sanders and Dave Martin of the Centre for Aboriginal Economic Policy Research at the Australian National University. Ian Heyward, also of the ANU, assisted with cartographic reproduction.
2 In Australian statistical geography, cities with more than 100,000 population are classified as major urban. In 2001, 60 per cent of the population lived in the five major cities of Sydney, Melbourne, Brisbane, Adelaide and Perth.
3 The band refers to a fluid residential camp or group of households that combine to occupy and subsist of a given range, or area of land. Typically, such groupings average 40–50 persons, though generally smaller in the arid zone.
4 In Australian Bureau of Statistics' parlance, discrete communities are defined as geographic locations that are bounded by physical or cadastral boundaries, and inhabited or intended to be inhabited predominantly by Indigenous people (more than 50 per cent), with housing and infrastructure that is either owned or managed on a community basis.

References

Altman, J.C. (1987) *Hunter-Gatherers Today: An Aboriginal Economy in North Australia*, Canberra: Australian Institute of Aboriginal Studies.

Altman, J.C. (2000) 'The economic status of Indigenous Australians', *CAEPR Discussion Paper No. 193*, Canberra: Centre for Aboriginal Economic Policy Research, The Australian National University.

Altman, J.C. (2001) *Sustainable development options on Aboriginal land: the hybrid economy in the twenty-first century*, CAEPR Discussion Paper No. 226, Canberra: Center for Aboriginal Economic Policy Research, The Australian National University.

Altman, J.C. and Nieuwenhuysen, J. (1979) *The Economic Status of Australian Aborigines*, Cambridge: Cambridge University Press.

Altman, J.C. and Sanders, W. (1995) 'From exclusion to dependence: Aborigines and the welfare state in Australia', in J. Dixon and R. Scheurell (eds) *Social Welfare for Indigenous Populations*, London: Routledge.

Altman, J.C. and Taylor, L. (1989) *The Economic Viability of Aboriginal Outstations and Homelands,* Report to the Australian Council for Employment and Training, Canberra: Australian Government Publishing Service.

Anderson, C. (1986) 'Queensland Aboriginal peoples today', in J.H. Holmes (ed.) *Queensland: A Geographical Interpretation*, Brisbane: Queensland Geographical Journal 4th Series.

Australian Bureau of Statistics (ABS) (1998) *Experimental Projections of the Aboriginal and Torres Strait Islander Population: 1996–2006*, Catalogue No. 2035.0, Canberra: ABS.

Baker, R. (1990) 'Coming in? The Yanyuwa as a case study in the geography of contact history', *Aboriginal History*, 14(1): 28–60.

Ball, R.E. (1985) 'The economic situation of Aborigines in Newcastle, 1982', *Australian Aboriginal Studies*, 1985/1: 2–21.

Barwick, D. (1962) 'Economic absorption without assimilation? the case of some Melbourne part-Aboriginal families', *Oceania*, 33(1): 18–23.

Barwick, D. (1964) 'The self-conscious people of Melbourne', in M. Reay (ed.) *Aborigines Now: New Perspectives in the Study of Aboriginal Communities*, Sydney: Angus and Robertson.

Beasley, P. (1970) 'The Aboriginal household in Sydney', in R. Taft, J. Dawson and P. Beasley (eds) *Attitudes and Social Conditions*, Canberra: Australian National University Press.

Beckett, J. (1965) 'Kinship, mobility and community among part-Aborigines in rural Australia', *International Journal of Comparative Sociology*, 6(1): 6–23.

Bell, M.J. (1992) *Internal Migration in Australia, 1981–1986*, Canberra: Australian Government Publishing Service.

Bell, M.J. (1995) *Internal Migration in Australia 1986–1991: Overview Report*, Canberra: Australian Government Publishing Service.

Bell, M. and Hugo, G.J. (2000) *Internal Migration in Australia 1991–96: Overview and the Overseas-born*, Canberra: Department of Immigration and Multicultural Affairs.

Bell, M.J. and Ward, G. (2000) 'Comparing temporary mobility with permanent migration', *Tourism Geographies*, 2(1): 87–107.

Birdsall, C. (1988) 'All one family', in I. Keen (ed.) *Being Black: Aboriginal Cultures in Settled Australia*, Canberra: Aboriginal Studies Press.

Borrie, W.D. (1994) 'Progress in Australian demography', *Journal of the Australian Population Association*, 11(1): 1–9.

Brady, M. (1987) 'Leaving the spinifex: the impact of rations, missions and the atomic tests on the southern Pitjantjatjara', *Records of the South Australian Museum*, 20: 35–45.

Brady, M. (1999) 'The politics of space and mobility: controlling the Ooldea/Yalata Aborigines, 1952–1982', *Aboriginal History*, 23: 1–23.

Brown, J.W., Hirschfeld, R. and Smith, D. (1974) *Aboriginals and Islanders in Brisbane*, Canberra: Australian Government Publishing Service.

Bryant, J. (1982) 'The Robinvale community', in E.A. Young and E.K. Fisk (eds) *Town Populations*, Canberra: Development Studies Centre, Australian National University.

Burnley, I.H. and Routh, N. (1985) 'Aboriginal migration to inner Sydney', in I.H. Burnley and J. Forrest (eds.) *Living in Cities: Urbanism and Society in Metropolitan Australia*, Sydney: Allen and Unwin.

Cane, S. and Stanley, O. (1985) *Land Use and Resources in Desert Homelands*, Darwin: North Australia Research Unit, Australian National University.

Castle, R.G. and Hagan, J.S. (1984) 'Aboriginal unemployment in rural New South Wales 1883–1982', in R.G. Castle and J. Mangan (eds) *Unemployment in the Eighties*, Melbourne: Longman Cheshire.

Commonwealth of Australia (1987) *Return to Country: The Aboriginal Homelands Movement in Australia*, Report of the House of Representatives Standing Committee on Aboriginal Affairs, Canberra: Australian Government Publishing Service.

Coombs, H.C. (1974) 'Decentralization trends among Aboriginal communities', *Search*, 5(4): 135–43.

Coombs, H.C., Dexter, B.G. and Hiatt, L.R (1982) 'The outstation movement in Aboriginal Australia', in E. Leacock and R.B. Lee (eds) *Politics and History in Band Societies*, Cambridge: Cambridge University Press.

Coulehan, K. (1995) 'Keeping company in sickness and in health: Yolngu from northeast Arnhem Land and medical related transience and migration to Darwin', in G. Robinson (ed.) *Aboriginal Health: Social and Cultural Transitions*, Darwin: NTU Press.

Drakakis-Smith, D. (1980) 'Alice through the looking glass: marginalisation in the Aboriginal town camps of Alice Springs', *Environment and Planning A*, 12: 427–48.

Drakakis-Smith, D. (1981) 'Aboriginal access to housing in Alice Springs', *Australian Geographer*, 15(1): 39–57.

Eschbach, K., Supple, K. and Snipp, M.C. (1998) 'Changes in racial identification and the educational attainment of American Indians, 1970–1990', *Demography*, 35(1): 35–43.

Gale, F. (1967) 'Patterns of post-European Aboriginal migration', *Proceedings of the Royal Geographical Society of Australasia, South Australian Branch*, 67: 21–38.

Gale, F. (1972) *Urban Aborigines*, Canberra: Australian National University Press.

Gale, F. and Wundersitz, J. (1982) *Adelaide Aborigines: A Case Study of Urban Life 1966–1981*, Canberra: Development Studies Centre, Australian National University.

Gray, A. (1985) 'Some myths in the demography of Aboriginal Australians', *Journal of the Australian Population Association*, 2(2): 136–49.

Gray, A. (1989) 'Aboriginal migration to the cities', *Journal of the Australian Population Association*, 6(2): 122–44.

Gray, A. (1997) 'The explosion of aboriginality: components of indigenous population growth 1991–96', *CAEPR Discussion Paper No. 142*, Canberra: Centre for Aboriginal Economic Policy Research, The Australian National University.

Gray, A. and Smith, L.R. (1983) 'The size of the Aboriginal population', *Australian Aboriginal Studies*, 1983(1): 2–9.

Hamilton, A. (1987) 'Coming and going: Aboriginal mobility in north-west South Australia 1970–71', *Records of the South Australian Museum*, 20: 47–57.

Holmes, J.H. (1988) 'Remote settlements', in R.L. Heathcote (ed.) *The Australian Experience: Essays in Australian Land Settlement and Resource Management*, Melbourne: Longman.

Holmes, J.H. (2002) 'Diversity and change in Australia's rangelands: a post-productivist transition with a difference?', *Transactions of the Institute of British Geographers*, 362–86.

Hugo, G. (1988) 'Population transitions in Australia', in R.L. Heathcote and J.A. Mabbutt (eds) *Land Water and People: Geographical Essays in Australian Resource Management*, Sydney: Allen and Unwin.

Inglis, J. (1964) 'Dispersal of Aboriginal families in South Australia (1860–1960)', in M. Reay (ed.) *Aborigines Now: New Perspectives in the Study of Aboriginal Communities*, Sydney: Angus and Robertson.

Lea, J.P. (1987) *Government and the Community in Katherine, 1937–78*, Darwin: North Australia Research Unit, Australian National University.

Lea, J.P. (1989) *Government and the Community in Tennant Creek, 1947–78*, Darwin: North Australia Research Unit, Australian National University.

Long, J.P.M. (1970) *Aboriginal Settlements: A Survey of Institutional Communities in Eastern Australia*, Canberra: Australian National University Press.

Long, J.P.M. (1989) 'Leaving the desert: actors and sufferers in the Aboriginal exodus from the western desert', *Aboriginal History*, 13(1): 9–43.

McKenzie, F. (1994) *Regional Population Decline in Australia*, Canberra: Australian Government Publishing Service.

Merlan, F. (1998) *Caging the Rainbow: Places, Politics and Aborigines in a North Australian Town*, Honolulu: University of Hawai'i Press.

Mitchell, I.S. and Cawte, J.E. (1977) 'The Aboriginal family voluntary resettlement scheme: an approach to Aboriginal adaptation', *Australian and New Zealand Journal of Psychiatry*, 11: 29–35.

Northern Territory Department of Education (1999) *Learning Lessons: An Independent Review of indigenous Education in the Northern Territory*, Darwin: Northern Territory Department of Education.

Palmer, K. and Brady, M. (1991) *Diet and Dust in the Desert: An Aboriginal Community, Maralinga Lands, South Australia*, Canberra: Aboriginal Studies Press.

Passel, J.S. (1996) 'The growing American Indian population, 1969–1990: beyond demography', in G.D. Sandefur, R.R. Rindfuss and B. Cohen (eds) *Changing Numbers, Changing Needs: American Indian Demography and Public Health*, Washington DC: National Academy Press.

Peterson, N. (in collaboration with J. Long) (1986) *Australian Territorial Organisation: A Band Perspective*, Sydney: Oceania Monograph 30.

Pollack, D.P. (2001) 'Indigenous land in Australia: a quantitative assessment of

Indigenous landholdings in 2000', *CAEPR Discussion Paper No. 221*, Canberra: Centre for Aboriginal Economic Policy Research, The Australian National University.

Povinelli, E.A. (1993) *Labor's Lot: The Power, History, and Culture of Aboriginal Action*, Chicago: University of Chicago Press.

Read, P. and Japaljarri, E.J. (1978) 'The price of tobacco: the journey of the Warmala to Wave Hill 1928', *Aboriginal History*, 2(2): 140–8.

Reynolds, H. (1987) *Frontier*, Sydney: Allen and Unwin.

Reynolds, H. (1992) *The Law of the Land*, Ringwood: Penguin Books.

Rowley, C.D. (1970) *The Destruction of Aboriginal Society*, Ringwood: Penguin Books.

Rowley C.D. (1971a) *Outcasts in White Australia*, Canberra: Australian National University Press.

Rowley, C.D. (1971b) *The Remote Aborigines*, Canberra: Australian National University Press.

Sahlins, M. (1999) 'What is anthropological enlightenment? some lessons of the twentieth century', *Annual Review of Anthropology*, 28(1): i–xxiii.

Sanders, W. (1999) *Unemployment Payments, the Activity Test, and Indigenous Australians: Understanding Breach Rates*, Research Monograph No. 15, Canberra: Centre for Aboriginal Economic Policy Research, The Australian National University.

Sansom, B. (1980) *The Camp at Wallaby Cross: Aboriginal Fringe-dwellers in Darwin*, Canberra: Australian Institute of Aboriginal Studies.

Sansom, B. (1982) 'The Aboriginal commonality', in R.M. Berndt (ed.) *Aboriginal Sites, Rights and Resource Development*, Nedlands: University of Western Australia Press.

Skeldon, R. (1990) *Population Mobility in Developing Countries: A Reinterpretation*, London: Bellhaven Press.

Skeldon, R. (1997) *Migration and Development: A Global Perspective*, London: Longman.

Smith, H.M. and Biddle, E.H. (1975) *Look Forward Not Back: Aborigines in Metropolitan Brisbane, 1965–1966*, Canberra: Australian National University Press.

Smith, L.R. (1980a) *The Aboriginal Population of Australia*, Canberra: Australian National University Press.

Smith, L.R. (1980b) 'New black town or black new town: the urbanization of Aborigines', in I.H. Burnley, R.J. Pryor and D.T. Rowland (eds) *Mobility and Community Change in Australia*, St Lucia: University of Queensland Press.

Taylor, J. (1986) 'Measuring circulation in Botswana', *Area*, 18(3): 203–8.

Taylor, J. (1989) 'Public policy and Aboriginal population mobility: insights from the Northern Territory', *Australian Geographer*, 20(1): 47–53.

Taylor, J. (1992) 'Geographic location and Aboriginal economic status: a census-based analysis of outstations in the Northern Territory', *Australian Geographical Studies*, 30(2): 163–84.

Taylor, J. (1997) 'Policy implications of Indigenous population change, 1991–1996', *People and Place*, 5(4): 1–10.

Taylor, J. (1998) 'Measuring short-term population mobility among indigenous Australians: options and implications', *Australian Geographer*, 29(1): 125–37.

Taylor, J. (2000) 'Transformations of the Indigenous population: recent and future trends', *CAEPR Discussion Paper No. 194*, Canberra: Centre for Aboriginal Economic Policy Research, The Australian National University.

Taylor, J. (2002) 'The spatial context of Indigenous service delivery', *CAEPR Working*

Paper No. 16, Canberra: Centre for Aboriginal Economic Policy Research, The Australian National University.

Taylor, J. and Arthur, W.S. (1993) 'Spatial redistribution of the Torres Strait Islander population: a preliminary analysis', *Australian Geographer*, 23(2): 26–39.

Taylor, J. and Bell, M. (1996) 'Mobility among indigenous Australians', in P.W. Newton and M. Bell (eds) *Population Shift: Mobility and Change in Australia*, Canberra: Australian Government Publishing Service.

Taylor, J and Bell, M. (1999) 'Changing places: Indigenous population movement in the 1990s', *CAEPR Discussion Paper No. 189*, Canberra: Centre for Aboriginal Economic Policy Research, Australian National University.

Uzzell, D. (1976) 'Ethnography of migration: breaking out of the bi-polar myth', in D. Guillet and D. Uzzell (eds) *New Approaches to the Study of Migration*, Houston: Rice University.

Young, E.A. (1981) 'The medium-sized town in the context of mobility', in G.W. Jones and H.V. Richter (eds) *Population Mobility and Development*, Canberra: Development Studies Centre, Australian National University.

Young, E.A. (1990) 'Aboriginal population mobility and service provisions: a framework for analysis', in B. Meehan and N. White (eds) *Hunter-Gatherer Demography: Past and Present*, Sydney: Oceania Monograph 39.

Young, E.A. and Doohan, K. (1989) *Mobility for Survival: A Process Analysis of Aboriginal Population Movement in Central Australia*, Darwin: North Australia Research Unit, Australian National University.

Zelinsky, W. (1971) 'The hypothesis of the mobility transition', *Geographical Review*, 61: 219–49.

3 Flirting with Zelinsky in Aotearoa/New Zealand

A Maori mobility transition

Richard Bedford and Ian Pool

There are definite patterned regularities in the growth of personal mobility through space-time during recent history, and these regularities comprise an essential component of the modernisation process.

(Zelinsky 1971: 221–2)

[D]espite the objections directed at transition theory itself, [the hypothesis of the mobility transition] has served, and continues to serve, as an adaptable and valuable framework for the investigation of population mobility.

(Skeldon 1997: 37)

While it may be optimistic, even perilous, to contemplate a global Indigenous variant of the mobility transition hypothesis, the degree of crossnational correspondence in mobility behaviour that appears to exist between Indigenous peoples cannot be overlooked.

(Taylor and Bell 1996: 164)

Context for a debate

Our chapter takes its lead from a seminal study by Ron Skeldon (1997). His book *Migration and Development: A Global Perspective,* devotes more of its review of theories and approaches to "transition theory," especially Zelinsky's classical mobility transition frame-work, than to any other theoretical perspective. Skeldon (1997: 37) argued, "transition theory . . . has served, and continues to serve, as an adaptable and valuable framework for the investigation of population mobility".[1] Similarly, Taylor and Bell (1996) articulated a clear case for an "Indigenous mobility transition". Using the example of Australia's Aborigines, they argued, *inter alia*, that a history of land dispossession, and the subsequent existence of Indigenous peoples on the margins of colonial and post-colonial settler societies, created a unique social and economic context for Indigenous mobility (Taylor and Bell 1996: 154). We thus ground our analysis of the geographical mobility of the Indigenous Maori population of Aotearoa (the islands that became New Zealand) in transition theory as it provides a useful framework for examining systemic shifts in both patterns of movement as well as the spatial distribution of Maori.

Structure of an analysis

Our argument comprises three sections. The first establishes the context for Maori mobility with reference to Zelinsky's (1971) central proposition: a functional relationship between the growth of personal mobility through time and across space and the "modernization process" associated with the urban-industrial transformation in all societies.

The first section also deals briefly with the problem of defining who is Maori. This is essential for understanding every aspect of New Zealand's social demography, including a Maori mobility transition (Metge 1971). Maori have intermarried overtly and legally with non-Maori (termed Pakeha, mainly European origin) since the earliest days of contact. While the Maori population today comprises under half who identify with more than one ethnic group, this is indicative of cultural affiliation, not descent.

A leading Maori scholar has argued that there are probably no Maori today who are biologically "full" Maori (Walker 1990a; see also Butterworth and Mako 1989), a situation that has held true for at least 40 years (Pool 1991: 22–5). More important, however, is that cultural identity has become the key determinant of ethnicity. It is thus necessary to establish just who we are considering to be Maori for the purposes of this discussion (elaborated in Kukutai 2001). This is a problem not just for New Zealand data, but is inherent in the analysis of Indigenous minority populations in Latin America, Anglo-America, the Arctic rim and Australia (Pool 1986).

The second section reviews the history of Maori mobility covering sequential phases commencing in pre-contact Maori society. For this we draw heavily on Pool's (1977, 1991) empirical studies of Maori population dynamics. This review is not a definitive study of Maori mobility, but it allows us to assess these broad phases against Zelinsky's (1971: 230–3) model, and to test Zelinsky's ideas using Maori mobility as a case study.

Mobility was one of a number of responses Maori made to disruptions following contact with and later colonization by Europeans. New Zealand's colonial and post-colonial history is being rewritten through research on these disruptions. This research is for judicial processes dealing with grievances coming from the Treaty of Waitangi, New Zealand's most fundamental constitutional document signed between Maori and the British Crown in February 1840. A significant part of these analyses relates to post-1840 movements of Maori following land alienation, both by *raupatu* (confiscation from those tribes who fought the Crown), and by land-court and other ostensibly legal processes that also had severe demographic impacts. We cannot examine in any detail research findings that have emerged in recent years (synthesized most recently in Belich 1996, 2001; but see, for example, Waitangi Tribunal 1997).

The final section returns to the theme of a possible generic Indigenous variant to Zelinsky's mobility transition. Clearly, the Indigenous minorities of North America, Australasia and elsewhere share common histories of displacement by white colonial settlement, the rise and subsequent abatement of post-war redistribution to urban centres, then an increased dispersal of population and settlement in rural areas. While reference to this common experience is useful, it should also not obscure other unique features of the Maori mobility transition, such as a convergence with trends in the mobility of the majority European descent (Pakeha) population.

For example, a critical component of Maori migration, shared with Pakeha since the 1970s, has been movement overseas, especially to Australia (Lowe 1990; Bedford 2001, writing on the New Zealand diaspora). In 1996, some 10 per cent of New Zealand's total population was living in Australia, as were roughly one-tenth of all Maori.[2] This convergence in patterns is partly a function of New Zealand's peculiar social demography: for example, through inter-marriage a migratory New Zealand family is often bi-cultural, Maori and Pakeha.

Situating a Maori mobility transition

The history of Maori mobility involves essentially a transformation from a completely rural society early in the nineteenth century to an overwhelmingly urban and cosmopolitan society in the late twentieth. The "urban transition", one of several socio-demographic changes associated with the "modernization" of societies in Europe from the sixteenth century, and later elsewhere, is an integral part of Zelinsky's (1971: 221–2) framework. Towns did not exist in pre-contact Aotearoa but owe their origin to European settlement from the early years of the nineteenth century.

This transition was far from a benign evolutionary process. Many of the social and economic transformations experienced over the past 200 years by Maori have been negative rather than positive resulting in under-development rather than development (see, for example, Walker 1996). In this regard, the Maori experience fits the general case Taylor and Bell (1996: 153) posited for Indigenous peoples in settler societies of the "New World".

The demography of Indigenous groups in North America and Australasia is different from that of Third World populations in significant ways, and is as much a manifestation of inequitable power relations and marginalization in the midst of plenty as it is to do with any lack of development *per se*.

Colonization produced an upheaval characterized by a systematic dispossession of Maori of their lands and the rapid inflow of Pakeha settlers (Pool 1991: 87–8). An "urban transition" became inevitable in the mid-twentieth century, when Maori owned only 5 per cent of New Zealand's total land area. Urbanization and the development of new systems of internal and international mobility were not options, but essential for the on-going survival of a population growing more rapidly by the 1940s than at any time in its recorded history (Pool 1991).

By 1996, 82 per cent of Maori lived in towns and cities with populations of 1,000 or more (New Zealand's definition of "urban"). Moreover, Maori mobility had become truly international. Around 40,000 Maori were living in Australia, mainly in its major metropolises. Unknown numbers live in the United States – many Maori Mormons have settled in Utah and Hawaii. Maori professional sports-people compete across the northern hemisphere; registered members of the South Island *iwi* (tribe), Ngai Tahu, are in Iceland, descendants of people who have been there for 200 years; and Diva Dame Kiri te Kanawa, is among London's large Maori community which has formed their own neo-*iwi*, Ngati Ranana.

Our argument about a Maori mobility transition addresses three substantive issues.

1   *Whether or not the Maori population was particularly mobile before World War II*. Maori share a tradition of mobility common to Polynesian societies renowned for their feats of oceanic navigation (Te Rangihiroa 1952; Lewis 1972; Hau'ofa 1994, 1998). From earliest contact with Europeans (late eighteenth century) some Maori have migrated overseas. Alongside international migration is the record of regional movements within New Zealand during the first 150 years of sustained contact with Europeans.

2   *The rapid growth in spatial mobility amongst Maori in the third quarter of the twentieth century*, particularly a rural exodus starting in World War II and reaching a peak in the 1950s to early 1960s, a period in which similar movements took place elsewhere overseas, including developed countries such as France. The feature standing out most for Maori is the sheer velocity of this movement: prior to the 1970s the most rapid on record for a national population according to a University of California, Berkeley, study (cited in Gibson 1973: 82). This urban movement had a major impact on Maori society, on *iwi* (tribes),[3] on *hapu* (sub-tribes) and on *whanau* (extended families). The Maori anthropologist Hopa (1996) has referred to this effect as the "torn *whariki*" (literally 'torn tissue').

3   *Whether or not Maori are at the "end" of their mobility transition now that they are essentially an urban population*. Here we recognize Zelinsky's point that the final stage of the mobility transition is not associated with low levels of mobility, thereby contrasting with the final stage of the classical demographic transition model defined by low levels of mortality and fertility. He argues that mobility of certain kinds increases at this stage, and we show similarly that Maori are no "less mobile" now that they are essentially urban.

Accompanying geographical movement are other forms of social mobility, most importantly flows in and out of the statistically defined Maori population. This is because "[a] feature of Maori–Pakeha relations is their fluid quality" (R.H.T. Thompson 1963: 54–6). Thus inter-ethnic mobility has methodological implications for our argument, and is discussed in the last part of this section.

Underlying theoretical issues

In all probability the Maori mobility transition is more "advanced", in terms of passing through phases identified in Zelinsky's classical transitional sequence, than is true for other Indigenous minority populations. Maori differ from Aborigines in their achievement both of a massive redistribution of the population from rural to urban places, as well as their diaspora overseas.

There are three theoretical issues that are particularly relevant in this context. Firstly, as we have noted, there is the "goodness of fit" between the Maori experience and Zelinsky's hypothesis and its revisions (Zelinsky 1971, 1979, 1983, 1993). Here we must note that his framework is merely one among a number of paradigms in the transition family of models analysing demographic changes. These include, for example, the classical demographic transition (Notestein 1945: the parent model), the industrial labour force transformation (Chenery and Syrquin 1975), and Omran's (1982) epidemiological transition. All such models share attributes: all attempt to synthesize and generalize from observed experiences across a wide range of societies, and to explain demographic phenomena by reference to wider aspects of social change. Thus Zelinsky's (1971: 221–2) schema delineates a general question relevant to Maori: are there "regularities in the growth of personal mobility through space-time during recent [Maori] history, and [do] these regularities comprise an essential component of the modernisation process?"

A second theoretical question relates to something plaguing all models of societal change, even those of the great social thinkers such as Marx. All start from some theoretical point, usually perceived as one of relative social stability, when societies are "pre-industrial" or "pre-modern". Zelinsky's transition commences with relatively immobile societies, although he subsequently acknowledged that "pre-modern" societies need not have low levels of spatial mobility (Zelinsky 1971, 1979, 1983). The frameworks then posit a terminal point, again often postulated as one of demographic stationarity and replacement,[4] or sometimes as the attainment of modernization. We ask whether Maori prior to World War II lived in a pre-industrial society with limited mobility.

Petersen (1961: 12) explicitly linked the processes of industrialization to the general demographic transition, reminding us how interrelated mobility is to other population changes, that together are also interlinked with social transformation in general. Skeldon (1997) emphasizes this point, noting that very few researchers have ever picked up the challenge to demonstrate, empirically, the mechanisms that link the different transitions associated with modernization. He is one of the few who has elaborated the linkages postulated by Zelinsky between the mobility and demographic transitions (Skeldon 1985, 1990, 1992, 1997; Zelinsky 1971, 1993).

Finally, we note, but only in passing, a point raised elsewhere by Pool (2000a, b) that some transition models have a distinct advantage over their parent. Zelinsky's considers not only the dynamics of mobility, but also its structural dimensions (e.g. the proportion of the population urban). Notestein's (1945) demographic transition only modelled shifts in natural increase, seeing structural effects of these as outcomes of changes in dynamics, but not an integral component of the transition itself. This has direct relevance for our chapter, relating to both dynamics and the structural dimensions of Maori mobility.

The Maori population: definitional issues

The exact number of Maori in New Zealand itself, let alone overseas, is an issue of dispute because of the difficulties inherent in defining Maori (Pool 1991: chapter 2 examines this significant question, having major policy implications). Demarcation is difficult because of the high levels of inter-marriage, both formal and informal, persisting from first contact with Europeans, linked to another phenomenon which this chapter cannot analyse, "inter-ethnic mobility" (Pool 1977: 46–8; updated in Brown 1983: appendix D). Category jumping occurs in both directions, with net changes and directions varying from one census to another, and between censuses and other data sources.

Therefore, New Zealand's ethnic mosaic is increasingly complex, but this has a beneficial impact. In the words of Ranginui Walker, "Maori and Pakeha are more like each other than they care to admit. There's a high degree of inter-marriage, which people forget is an important part of integration in this country" (Walker 2002: A16).

Until and including the 1981 census, the Maori population comprised anyone reporting themselves as "half or more" Maori. But it had long been clear (Pool 1963) that this definition was not only inappropriate in such an ethnically fluid situation, but was essentially bypassed by Maori who reported how they felt (Pool 1991: 22–5). Since 1986 the census has better reflected reality. Indeed, New Zealand became one of the first countries, probably the first, to allow multiple responses to the ethnic identity question, with no attempt to get respondents to rank these. This fits well with socio-cultural norms but poses major problems for statisticians resolved arbitrarily, and rather inelegantly, by classifying everyone who reports that they are "Maori", or "Maori" and something else, ethnically as "Maori". The population as defined, and referred to here, is probably larger than that whose prime identity is really Maori. This is confirmed by small-scale surveys (e.g. on hospital patients reported Pool 1991: 15), and in a nation-wide sample survey of 3,000 New Zealand women aged 20–59 years on fertility and family formation directed by Pool (Pool *et al.* 1999; Marsault *et al.* 1997; Kukutai 2001).

The importance of this for our chapter is threefold. First, it deals with a very fluid social system, but where fluidity manifests itself overtly and legitimately. Second, Maori are involved in several sorts of mobility, only one of which we are addressing here. Third, we cannot be certain how many "Maori" there are, but at the 1996 census 523,000 persons reported Maori ethnicity, representing 14.5 per cent of the total population, a higher proportion than is true for Australian and North American Indigenous minorities. This factor is central to New Zealand's social demography.

A Maori mobility transition

There are no written records of the Maori occupation of Aotearoa before the arrival of Europeans. Sources remain oral traditions, including the post-contact European and Maori transcriptions of these, and archaeology. We present here merely a generalized overview of Maori mobility in the "classical period" of their history, mainly to establish whether Maori mobility conforms with Zelinsky's general conclusions about mobility in "traditional society[ies]". This is followed by an overview of Maori social and economic transformation since the signing of the Treaty of Waitangi in 1840 with reference to four key periods: colonial rule, 1840–1906; the tranquil years, 1906–1945; Maori urbanization, 1945–1971; Maori mobility and globalization, 1971–2001. The migration patterns characterizing these periods are assessed alongside Zelinsky's propositions about mobility for each of the sequences he identifies.

Classical period: prehistory to 1840

While the timing of Maori arrival in New Zealand is disputed among prehistorians it seems that this migration was recent, say 800–1,000+ years ago. Also controversial is whether or not return migrations by Maori to the Pacific Islands occurred in prehistory. Te Rangihiroa (1952), who carried out many of the baseline Pacific ethnographies, called Polynesians "Vikings of the Sunrise" in recognition of their prowess as ocean navigators.

Maori spread throughout New Zealand, but were concentrated in the northern third where the tropical crops they introduced could survive. Following "high island" Pacific traditions, Maori subsistence was dependent on hunting, extensive fishing and gathering, and the production of root crops (Belich 1996: 67–75). The primary production unit was the family (*whanau*), grouped into *hapu* that were, arguably, the primary units dealing with the allocation of land. *Hapu*, in turn, clustered into *iwi* (tribes), which were the cardinal political groupings, often linked back through genealogy to broader entities, each descended from particular canoes that reached Aotearoa (Belich 1996: 83–6).[5]

There was extensive social and economic interaction between Maori groups. Until over-hunting rendered extinct the giant flightless bird, the *moa,* there was a trade in preserved meat from the South Island, where moa were abundant, to the concentrations of Maori in the north. Other south-to-north trade was in jade from the southwest, and in other goods.

Belich (1996: 34) has argued that moa-hunting created a "protein boom", so that its extinction profoundly affected the economy and thus mobility. Competition for resources led to widespread warfare, and significant movements of conquest. The evidence for this comes from *whakapapa* (oral, mainly genealogical, history), and the remarkable number of *pa*-sites (fortified settlements), dotting hillsides especially in the north where the pressures of population would have been most intense (Stone 2001).

Wars and population displacement may have become more significant after contact with Europeans when muskets became available especially to coastal tribes. The "musket wars" of the 1820s and 1930s unleashed movements that abated only in the late 1830s "when New Zealand returned slowly to a balance of arms such as it had enjoyed before European weapons were introduced" (Wright 1959: 101). As Vayda (1960: 85) argued, the number of persons killed, at least directly, "in old time fights are obvious exaggerations . . . " but it is certain that these wars set in train massive migrations. This is shown by the painstaking analysis of Urlich (1969). She sifted through and ordered population data by region and tribe, then Pool (1991: chapter 3) compared these distributions with Dieffenbach's estimates (1840), Fenton's census (1857) and the 1874 census.

Several key points stand out. First, in major northern areas, such as around Auckland and the Bay of Plenty, large populations were displaced. Many fled from there into the inland Waikato, and across to the eastern littoral. Some Taranaki Maori, displaced by Waikato *iwi,* moved southwards into the South Island, and then to the Chatham Islands where they decimated the Indigenous Moriori.[6] Second, in terms of their magnitude and long-term redistribution effects, these migrations probably exceeded all subsequent movements until the rural exodus after World War II. The wave-effects and counter-flows of these conflict-related migrations continued for decades, obscured, however, by other events: notably the Treaty of Waitangi (1840) and the New Zealand Wars (1840s and 1860s) between, on one side, some Maori, and on the other the Crown, Pakeha colonists and other Maori.

Links with Zelinsky

Pre-1840 Maori society resembles "phase 1" of Zelinsky's mobility transition, "pre-modern traditional society". A key difference after 1800 was the diffusion of muskets through pre-colonial Aotearoa, an innovation that transformed elements of Maori mobility and geography. Prior to the musket era mobility patterns had been similar to those found elsewhere in Oceania: that is, significant interaction between social groups and quite extensive residential as well as circular mobility (Bedford 1973). These were not what Zelinsky (1971: 234–5) termed "spatially stable peasant societies" where mobility was localized territorially.

While the primary social units could be characterized as "an array of cells firmly fixed in space . . . " (Zelinsky 1971: 234), there was considerable spatial mobility in "traditional" Maori society. This was central to a dynamic political economy in which changing environmental, social and economic conditions encouraged regrouping and reconfiguring, rather than long-term spatial and social stability.

For Zelinsky (1971: 234) "the migratory practices of hunting-and-gathering folk or cultivators at the tribal level have been so diverse that generalization seems futile". Moreover, there was no theoretical necessity to postulate "limited" mobility in all pre-modern, traditional societies (Zelinsky 1979, 1983). The *incidence* of mobility was not the critical issue, but rather its spatial forms and impacts on the redistribution of population from "traditional communities" to other kinds of settlements. For Maori the "early transitional phase", prior to the Treaty of Waitangi, was associated with an initial widening of options for mobility, including movement to coastal settlements that were springing up, and some overseas trade with Australia.

Colonial rule, 1840–1906

The Treaty of Waitangi was often honoured in the breach. Maori who fought the Crown lost land by confiscation, while all Maori, even "friendly", saw their land alienated by purchase through Land Court processes that were typically unethical, even illegal, charades. In the 1840s, there was scattered warfare, but the most protracted hostilities occurred in the 1860s. Conflicts over land acquisition for European settlers caused hardship and displacements especially in the central North Island (Waikato, Bay of Plenty, Poverty Bay, Taranaki and Wanganui: see Belich 1986).

The wars and legislation relating to Maori land ownership effectively destroyed an emerging Maori cash economy. From the 1840s to the 1860s Maori traders had not only come into the nascent colonial towns but had also exported goods to Australia and perhaps further afield (Belich 1996: 215). The Land Wars in parts of New Zealand in the 1860s saw land confiscated in some regions while the Maori Land Court, introduced in 1862 to arrange transfers of land between Maori and Crown agents, instituted a long-term process of alienation. The Crown individualized communally owned Maori land, and Crown agents inveigled or forced individuals to sell this. Land was also forcibly taken under Public Works and similar Acts for railway and other infrastructural development. There were, however, almost no reserves, so that the herding of people onto land set aside for "natives", as in other settler societies, did not occur in New Zealand. Instead, Maori retained smaller blocks of land dispersed between European colonists' farms or settlements.

All these land alienation processes forced Maori to move, but generally this involved local rather than long-distance displacements. Moreover, Court hearings had to be attended by entire *hapu*, even infants who otherwise forfeited title, forcing short-term circulation, often over long distances, but not significant long-term movements. The main impact of court hearings, lasting week after week with Maori congregated into un-hygienic camps near the courts, was on health and thus mortality rather than on population movement and redistribution (see Sorrenson's (1956) seminal study, substantiated by Pool (1991: chapters 4 and 5), and detailed for the Central North Island by unpublished work undertaken by Pool, T. Kukutai and J. Sceats for the Crown Forestry Rental Trust (CFRT), who have given permission to cite the study.

Essentially, Maori had to shift radically from extensive land-use to intensive subsistence. Initially this produced under-development, and, related to this, increased infant and childhood mortality (Pool 1991: chapters 4 and 5). As a survival strategy Maori sought alternatives to farming, such as casual wage employment, by travelling to areas where there were exploitative industries, particularly timber milling and digging kauri gum (resin of dead *Agathis australis* trees). Circular migration within rural areas, to European resource and settlement frontiers, set the scene for patterns of mobility that continued well into the twentieth century.

There were, however, positive measures. The 1877 Education Act, successfully implemented by about 1900, ensured access of Maori (and Pakeha) to free universal and compulsory education. Maori could attend the same schools as Pakeha and followed the same curriculum, although "native schools" in areas where Maori were concentrated accorded token recognition to Maori culture. A successful primary health campaign in the early 1900s was implemented by Maori medical practitioners within the newly formed Health Department. By 1911 Maori life expectation had increased rapidly to about the same level as that of Spain, but was still far below the Pakeha level (Pool 1991: 77).

Links with Zelinsky

The colonial period in New Zealand saw the destruction of the "pre-modern, traditional" Maori society and economy. But "early transitional" Maori society did not follow Zelinsky's (1971: 230) model that posited a rural exodus or "major outflows" to international destinations. Simultaneously, there was "significant immigration of [non-Maori] skilled workers, technicians, and professionals from more advanced parts of the world" (Pool 1991: figure 5.2),[7] and some movement of Maori to "rural

[European] colonization frontiers". The available evidence also shows the "growth in certain kinds of circulation" (Zelinsky 1971: 230), but this was mainly within rural areas, and was linked to the loss of land by Maori and disruption of their economy.

Zelinsky (1971: 236) associated modernization with a "great shaking loose of migrants from the countryside". Co-variants included shifts in replacement rates, "a general rise in material welfare . . . and improvements in transport and communications". Generally these changes were not achieved for Maori during the period of colonial rule between 1840 and 1906.

There was a major change in levels of replacement, but inversely to that hypothesized in the classical demographic transition model. Until the 1890s death rates remained high, and the population decreased in size. Maori numbered at most around 100,000 in the 1760s; by 1840 a "best" estimate is around 80,000; and at the 1891 census they numbered just under 42,000 (Pool 1991: 57, 76). Other than through education and limited health programmes, until the twentieth century, Maori material welfare was not addressed seriously by government.

Thus the second phase of the Maori mobility transition does not fit Zelinsky's "early transitional phase", even though colonialization had transformed Maori society. The most significant form of mobility became circular labour migration – a feature characteristic of Pacific colonial territories throughout the nineteenth century (Bedford 1973). Over the next 40 years, after New Zealand became a British Dominion, the rhetoric of Maori leaders and politicians favoured rural development and practical steps were taken to make farming more productive. The "shaking loose of migrants from the countryside" was not to occur until the irruption caused by World War II.

The tranquil years, 1906–1945

In 1906, New Zealand became a Dominion. Maori played a role in Dominion politics as they had even in the latter colonial years.[8] Notwithstanding this legal equality (Metge 1971: 55–6) and continuing interpersonal contacts (e.g. through sports team and legal marriage) Maori remained essentially marginal to much of mainstream New Zealand: geographically, socially, demographically, and economically (B. Thompson 1985: 124). This was a period in which Maori underwent "recuperation in isolation" (Pool 1991: chapter 6). In a sense, these were tranquil years after the turbulence of the colonial era: gradual demographic recovery and some economic progress as Maori leaders sought to regain control over the remaining Maori-owned resource base.

In the period 1906–1945, long-distance mobility was probably more limited than was the case a century earlier. Supporting this is a study of interprovincial migration using census survival methods (Brosnan 1986).

Prior to the first modern enumeration (1926) completeness varied from census to census, and from region to region, while age-reporting was very poor, producing age- and cohort-distortions that had critical effects on census survival ratios. After 1926, by contrast, data can be accepted with some degree of confidence although even in 1926 there may have been 10 per cent under-enumeration (Pool 1991: 106; 1977: chapter 3, appendix B).

Between 1926 and 1945 the average annual intercensal rates of Maori migration were between 1.0 and 1.4 per cent – well below Pakeha rates, except in 1936–1945 (Brosnan 1986: table 5). Against this there was undoubtedly circular mobility along with seasonal movements to rural industries associated with sheep- and beef-farming, especially to work shearing or in abattoirs. These conclusions are supported by a recent central North Island study (CFRT see above) employing census survival rates for the period 1926–1945. It shows that out-migration from this region, which has heavy Maori population concentrations, was limited except for men at the working ages.

During the 1920s and 1930s there were numerous attempts to develop a commercial base to Maori farming led by a Maori Cabinet Minister Sir Apirana Ngata, who argued that "the future of Maori 'lay principally in the land'" (cited by McCreary 1972: 193–4). Two types of rural commercial enterprises became dominant: consolidated holdings ("incorporations") farmed on behalf of groups of owners by a manager, and dispersed single units modelled on the dairy farms common amongst Pakeha farmers. The former did not always provide much employment for local Maori, while the latter were often too small and too isolated to be effective economically.

A combination of limited land resources, a gradual demographic recovery, the completion of many land development programmes, and the modernization of agriculture created labour pressures in rural areas where Maori were concentrated. By the 1950s, the policy of encouraging Maori to remain in rural areas in order to preserve their cultural identity and improve their economic prosperity through farming was no longer seen as being viable.

Between 1926 and 1936 the Maori population increased by more than 20 per cent[9] while the non-Maori population grew by only 11 per cent. The 1936 census highlighted differences between the younger age structure and rapid growth prospects for Maori as against the majority Pakeha whose structures and growth were affected by declines in fertility and immigration (Pool and Bedford 1997). Hidden underemployment and unemployment, and the limited opportunities for deriving a livelihood in rural areas forced increasing numbers of Maori to seek work elsewhere. During the war years Maori migration was significantly influenced by volunteering for military service and by recruitment into wartime industries.[10]

World War II was responsible for the first real exposure to urban living for many Maori. Moreover, there were flow-on effects: "When the wartime decentralization of industry failed to extend materially the limited range of employment opportunities to be found in those areas where Maori were most numerous, large numbers moved of their own accord to the chief towns and cities" (Metge 1952: 112).

By 1945 the Maori population had returned to the size it had been 100 years earlier. Whereas in 1840 they had owned over 90 per cent of the land, in 1945 this was only 5 per cent. The demographic conditions for "a great shaking loose of migrants from the countryside" (Zelinsky 1971: 236) had been realized for Maori after over a century of protracted contact with the forces of "modernization".

Links with Zelinsky

Maori continued to lose land though the period 1906–1945 while their numbers were recuperating after decades of decline, yet they remained essentially marginalized from the wider society and economy. By 1926 the majority of New Zealanders were urban, but most Maori remained rural. This was not due to policies fostering segregation. There were no formal reservations for Maori in New Zealand, and interpersonal relations between Maori and Pakeha had been forged by intermarriage for over a century. At a macro-level, however, New Zealand was essentially two separate geographic entities – Maori in isolated rural northern and eastern North Island areas, and Pakeha in cities and their accessible rural hinterlands. As we have noted, this *de facto* separation had been encouraged to a degree by Maori leaders promoting rural development.

Until the wartime urbanward movement, Maori fitted somewhere between Zelinsky's two phases of a "pre-modern traditional society" and an "early transitional society". Thus, their basic distribution patterns resembled what had prevailed in the colonial period. Circular forms of population movement predominated, particularly of labour in rural areas. Long-distance internal migration and even overseas movements were less significant than they had been 100 years earlier. In a sense an "infinite pause" (Howlett 1973) had settled over rural Maori society, only to be terminated by the war in the 1940s.

Maori urbanization, 1945–1971

Measurement of Maori urbanization is complicated by two definitional problems: first, as we have stressed, defining who is "Maori" and second, classifying what is "urban". Taking census definitions extant in 1936, only 7 per cent of Maori were living in cities and boroughs (Bedford and Heenan 1987: 143). To measure geographical redistribution of Maori over the following decades we employ a time series (1926–1976) constructed by Watson (1985: 122) of Maori rural and urban populations, using the definition of Maori employed prior to 1981 and consistent definitions of urban and rural.

At the 1945 census about a quarter of Maori (25,000) were urban; by 1956 this was 35 per cent; in 1966, when this population reached 200,000, a number (123,774) exceeding that estimated at contact, was living in towns and cities and constituted 62 per cent of all Maori (Watson 1985: 122). Five years later, the share in urban places had risen to 71 per cent – approaching the 78 per cent of the total population classed as urban.

Between 1945 and 1971 the rural exodus dominated all aspects of Maori social life obscuring three other contemporary major demographic changes. Two of these were related to, and determined by, migration: a major shift in the structure of the Maori industrial labour force, and significant changes in the dynamics of Maori families and social groups. The third, radical improvements in survivorship, came primarily from positive health and social policy interventions (Pool 1991: 147–51), and to a lesser degree from better access to health facilities in urban settings. Maori also benefited from improved housing in urban areas, because the urbanward movement had been encouraged by government (see below), in policies incorporating strategies to eliminate differences between Maori and Pakeha (Hunn 1961: 14).

Maori were given priority for state housing, as were low-income families and those exposed to tuberculosis (Maori were over-represented in these categories). They were treated exactly the same as Pakeha in New Zealand's comprehensive social policy system initiated in 1938, constituting a Scandinavian-style welfare state. Beyond this, house ownership was encouraged through favourably structured loans, and young Maori were encouraged into apprenticeships in the building trades. A strategy of "pepper-potting", or scattering state houses and their low-income occupants through residential neighbourhoods, was adopted in the post-war period, as an attempt, through housing policy, to ensure that ghettos did not develop (Metge 1971: chapter 14). Among other effects this changed the flows from being dominated mainly by adolescents and young adults towards family migration (Pool 1991; Metge 1995).

Ironically, although designed to facilitate integration of Maori into urban neighbourhoods, "pepper-potting" was subsequently attacked with some justification as being "assimilationist" (Pool 1977: 225, citing Ranginui Walker). In fact, this forward-thinking housing policy was eroded by the development of large, low-income housing estates on the peripheries of Auckland and Wellington especially – estates which became the destinations for increasing numbers of Maori and Pacific Island Polynesian immigrants during the 1960s and 1970s. By the 1970s, New Zealand's two major cities had sizeable "ethnic" suburbs in Otara-Mangere (Auckland) and Porirua (Wellington).

Successive post-war governments also actively encouraged movement of Maori to areas with potential for economic development (Poulsen *et al.* 1975: 323; Metge 1964; Pool 1977). Moreover, government proactively linked internal population redistribution to policies relating to the immigration of foreign-born – this was a period in which such inflows were at high levels (Farmer 1985; Bedford and Heenan 1987). This was summarized in a paper at an economic development conference in 1960, by a senior civil servant who observed: "We would be committing a supreme folly if we pursued overseas migration to the point of creating a racial [sic] problem of under-employed Maori pent up in 'Maori' areas while immigrant labour filled the vacancies for labour in other areas" (Woods 1960: 2).

The net result of all these initiatives, coupled with the rapid Maori population growth that gathered momentum during the post-war period, was a massive movement from the more isolated, rural areas of the north, east, central west and plateau regions of the North Island. Favoured destinations were neighbouring towns and small cities, Auckland and Wellington urban areas and, to a lesser degree, the other two major centres of that time, Christchurch and Dunedin. In 1945–1966, some smaller urban areas adjacent to regions with high concentrations of rural Maori recorded Maori population growth rates exceeding 200 per cent (Hamilton's was 828 per cent).

The urbanward movements were very significant, constituting in the peak decade (1956–1966) a figure equivalent to 68 per cent of the Maori population total. Between 1961 and 1966, when the share of the Maori population living in urban places increased from 54 to 62 per cent, the number of rural Maori actually declined by 14 per cent (Watson 1985: 122). In 1966 the Maori rural population of 77,321 was only 4,000 more than in 1945. The urban population, by contrast, had increased almost fivefold (Watson 1985: 122).

As noted earlier, at first post-war migrants to urban areas were concentrated at the youth ages.

> From 1956 to 1961, 18 per cent of all Maori aged 15–19 years and 11 per cent aged 20–24 years were net movers (i.e. this figure does not cover gross mobility, . . .) to the major urban centres alone. By 1961, 22 per cent of the Maori urban population was composed of net in-migrants of the period 1956–61 while this was true of 51 per cent of urban Maori aged 15–19 years.
>
> (Pool 1991: 153–4)

This pronounced age-specificity had a major impact on social organization both at origin and destination.

Other population movements included seasonal, circular and various shorter-term movements. Recalling that Maori internal migration was typically to neighbouring areas rather than longer distance, it is salient that in 1956–1961 even Maori interprovincial mobility rates exceeded those of Pakeha (Brosnan 1986). There were also major inter-regional shifts: between 1951 and 1956 just over 12 per cent of Maori moved, but at its peak in the two quinquennia 1956–1961 and 1961–1966 inter-regional migration involved 16–18 per cent of all Maori (Poulson and Johnston 1973).

Beyond this, there were major intra-urban migrations that altered the social ecology of New Zealand cities, especially Auckland and Wellington. At the end of the war and in the 1950s Maori in-migrants had been clustered in inner city areas, but there was then a shift, particularly of young families, to outer suburbs, typically the new low-income housing estates. By 1956, 47 per cent of outer suburban Maori were aged 0–14 years (a figure close to the national Maori level) as against only 33 per cent in the inner city (Pool 1961; Rowland 1971b).

In the 1960s and 1970s Maori urbanization was a topic attracting considerable attention from academics and policy makers (see Bedford and Heenan (1987: 152–5) for a review of these studies). Urbanization was leading to "the development of a group who, for want of a better name, might be called the new '*tangata whenua*', the people whose true 'home' is Auckland" (Pool 1961: 66). This did not mean that Maori in-migrants integrated fully into Pakeha urban society. Most migrants "reacted . . . not by eliminating the points of difference as quickly as possible, but by placing an increased value upon the emotional security provided by the Maori group" (Metge 1964: 263).

Such security was clearly desirable given the conclusions drawn by Rowland (1971a, b, 1972, 1973, 1974) and other researchers about the experiences of Maori in cities in the early 1970s. Rowland (1972: 20), for example, observed that "urbanisation had not enabled many Maori to achieve material success comparable with that of the average European" and that "a ceiling on progress is imposed by the fact that most Maoris cannot afford to live anywhere but in areas of cheap housing".

A major consequence of migration during this period was a Maori industrial labour force transformation. Maori men moved from the primary sector into manufacturing, and by 1966 this was the modal group, exceeding primary and tertiary combined. Maori women in remunerated employment had long been in primary and tertiary jobs, but by 1966 manufacturing was increasingly an alternative form of employment. Maori household incomes increased, while from the 1960s fertility decreased, so that Maori families became more prosperous (Pool 1991: chapters 7 and 8). But in the longer run this was to prove somewhat illusory as employment in manufacturing and construction industries into which Maori went, often as semi-skilled workers, suffered far more than other sectors in the radical economic restructuring in the late 1980s.

Maori social organization underwent radical shifts. The support networks binding together *whanua*, *hapu* and *iwi* were challenged by the clustering together of Maori from different areas in low-income housing estates. Competition for unskilled and semi-skilled jobs from immigrant Pacific Island peoples created tensions between *tangata whenua* and their "cousins" from island Polynesia (Fleras and Spoonley 1999).

Not surprisingly, this created less than ideal conditions for Maori by comparison with Pakeha and challenged conventional wisdom about ethnic harmony. Although on an international scale alongside other Indigenous minorities (Geddes 1961), Maori were relatively better off, they did face what today would be called vulnerability (Rowland 1972: 20; see also R.H.T. Thompson 1963; Harré 1963; Metge 1971: chapter 14). The situation was summarized perhaps most succinctly and fairly by the novelist Pearson (1962: 148): "The most striking feature of the Maori situation 17 years after the end of the [World] War [II] is the continued existence within the Welfare State of rural enclaves of material poverty and, in city and country, spiritual insecurity". Moreover, as Douglas (1979: 109) observed at the end of the 1970s, the social calculus had changed. Aspiring to a Pakeha lifestyle was no longer a key issue in Maori urbanization, as Maori wished to see themselves as an identifiable viable community with the " . . . right to be different and that the degree of difference is for the Maori themselves to determine . . . ".

The post-war period of mass Maori migration thus had contradictory effects. On the one hand it saw Maori escape from the marginalized situation they were in at least until World War II. Accompanying the shift-share in the labour force over the period 1951–1976, Maori incomes in general, and those for the actively engaged, converged markedly towards those of Pakeha (Martin 1998: 104).

Unfortunately, there are few time series for socio-economic data available sub-nationally, except census information on key housing amenities, piped water (as against rain tanks or none) and flush toilets. In 1945, almost all Pakeha houses in urban areas had these, and access to water and toilets for urban Maori was 91 and 77 per cent respectively of the Pakeha rate. In rural areas, where Pakeha ownership of these amenities was still not universal (53 per cent piped water; 32 per cent flush toilets), Maori levels were far lower (28 and 15 per cent respectively). But in 1945, 55 per cent of inhabited, private, permanent Pakeha homes were urban, whereas only 8 per cent of Maori were. Besides this, Maori levels of ownership of rural "temporary" dwellings, some 30 per cent of the total, far exceeded Pakeha rates (3 per cent nationally, CFRT study). By contrast, in 1971 when 51 per cent of Maori and 70 per cent of Pakeha houses were urban, piped water and flush toilets were almost universal in urban areas for both populations (98+ per cent). Most (74+ per cent) rural houses also had these amenities, but Maori rates were only 91 (piped water) and 82 (flush toilets) per cent of Pakeha.

On the other hand, the rural exodus wrenched apart Maori social structures and put them into a situation of economic vulnerability. The significant state-led economic restructuring in the late 1980s (see next section) was to have a profound effect economically on Maori in urban and rural areas. Their capacity to absorb such a shock had been lessened by the weakening of family and other structures in the period of mass rural exodus.

Parallel to these economic transformations was a Maori cultural revolution that in one sense challenged the extant philosophy of integration (Hunn 1961: 14). At the same time *de facto* integration through high levels of intermarriage was accelerating (Harré 1966) so that, at a macro-level, one could still see two separate cultures, to a degree in conflict. Yet at a micro-level many families were not only bicultural but interacting fully in the wider society.

Convergence in patterns of population distribution between Maori and Pakeha continued after 1971, although there were differences in the processes driving these movements. By 1971 Maori were into the third phase of Zelinsky's (1971) mobility transition – the "late transitional society". The "massive movement from countryside to cities", the hallmark of his "early transitional society" phase had, to all intents and purposes, ended for Maori by the early 1970s.

Links with Zelinsky

The post-war urbanization of Maori fits well with the second phase of Zelinsky's mobility transition, "a great shaking loose of migrants from the countryside" occurring between 1945 and 1971. This "new Maori migration", Metge's (1964) term for the rural exodus, co-varied in time with rapid declines in mortality that were followed by even more rapid decreases of fertility in the 1970s, thus changing the Maori "reproductive budget" (sic, Zelinsky 1971), as well as with "a general rise in material welfare or expectations and improvements in transport and communications". However, it is clear that this phase was not associated, temporally at least, with the onset of modernization, for Maori had been exposed to European influences for more than a century before the massive redistribution occurred.[11] Until World War II they had exercised other options identified by Zelinsky, but posited by him to be less likely. Until then most Maori had "neither the inclination nor the opportunity to desert the rural locality" (Zelinsky 1971: 236), at least in a more permanent way.

When Maori rural–urban migration finally "took off" dramatically it was triggered by a combination of state-led policies and socio-demographic pressures in Maori rural communities. In 25 years Maori shifted from being 75 per cent rural to just under 75 per cent urban. Not surprisingly, the social consequences of this revolution had severe impacts on both rural as well as urban communities, while nuclear and extended families, and even sub-tribes, were subject to extreme strain (Hopa's metaphor of the torn *Whariki*). In the 1960s the full implications of this social dislocation were little understood and perhaps did not become evident until identified in the 1990s (e.g. Hopa 1996, elaborating Pool 1991; Fleras and Spoonley 1999).

By the early 1970s there were signs of resistance, however, as manifested by protest by Maori over the loss of their lands, their language and their *mana* (authority) indicative that a significant cultural revival was in the making. This revival did not reverse Maori urbanization, which continued through the 1970s and 1980s so that by 1991, 82 per cent of Maori were urban residents. However, it did affect significantly the way Maori expressed their identity and strove to establish a distinctive place for their culture and values in cities.

Zelinsky's "early transitional society" is not only characterized by massive rural–urban migration within countries, but is associated also with "major outflows of emigrants to . . . foreign destinations", "movement of rural folk to colonisation frontiers", and "growth in various kinds of circulation". Maori did not experience substantial emigration to foreign destinations at this time but many did move to large-scale energy projects and newly developed forestry industries in what could be called "colonization frontiers" within New Zealand (e.g. Central North Island, CFRT see above). The rural "colonization frontier", in the sense of farming after World War II had been reduced to very small pockets of Maori land in most parts of the country. Towns, energy projects and forestry industries rather than "land suitable for pioneering" became the "frontier" for post-war Maori seeking a new life.

Maori migration to Australia and other overseas destinations had occurred since the beginning of the nineteenth century, but the numbers resident overseas were still believed to be relatively small. In 1966, there were 862 Maori in Australia (employing a definition close to that used in New Zealand at that time), "or perhaps three times this number if persons of Maori descent had been identified as well" (Pool 1991: 179). Twenty years later over 25,000 persons of Maori descent were enumerated in Australia (Lowe 1990). The Maori diaspora thus followed urbanization, and was not in tandem with it as suggested by Zelinsky (1971: 233).

The limited literature on patterns of Maori population circulation during the post-war period suggests that the drift to towns was not a one-way relocation. Auckland Maori knew that if "worst came to worst, they could always go home" (Metge, cited in Scott and Kearns 2000: 24), where they have their *turangawaewae* (places to stand). Circulation of Maori was driven by similar socio-cultural and economic imperatives that encouraged Pacific Islanders to maintain a bilocal pattern of residence, rather than relocating permanently in either a local urban destination or in one of New Zealand's cities (Bedford 2000). Polynesian migration to Auckland and Wellington especially, was gaining momentum in the 1960s and over-lapping with the influx of Maori into the cities (Bedford 1999).

By the early 1970s, Maori had crossed critical thresholds in both their mobility and their vital (classical demographic) transitions. Maori became a "late transitional society" in the 1970s with a co-varying "major decline in fertility" and a "slackening, but still major, movement from the countryside to the city" (Zelinsky 1971: 230). Maori "fertility declines were most rapid in the 1970s, dropping from 5.0 live births per woman in 1971 to 3.0 in

1976 . . . [while] by 1983, Maori fertility was close to the Pakeha level (Maori Total Fertility Rate=2.2, as against 1.9), and in fact Maori age-specific rates were lower at older ages" (Pool 1991: 166).[12] In the case of mobility, movement into larger cities, especially Auckland, continued, although at a slower rate than in the 1950s and 1960s (Pool 1991; Poot 1984). This presaged the return migration of Maori to rural communities and the movement across the Tasman that were to attract attention from researchers in the late 1970s.

Maori mobility in the era of globalization, 1971–2001

In the 1970s:

> Maori [were] generally less mobile than non-Maoris. Although Maori [had] a slightly higher rate of intra-urban mobility this [was] due to the relative youthful-ness of the Maori population. The propensity to move north and the urbanisation process [were] both stronger for Maoris than non-Maoris, and the redistribution of population through migration more unbalanced.
>
> (Poot 1984: 303)

By the 1980s and 1990s, some of these trends were changing. Maori were already highly urbanized so that their mobility and distribution patterns generally resembled those of Pakeha, and both populations, especially Maori, seemed to be highly mobile. At the 1996 census, 55 per cent of Maori had reported that they had lived somewhere else five years earlier (for Pakeha the figure was lower but still 50 per cent). Maori were most likely to have shifted locally (intra-urban mobility was high) or from else-where in New Zealand. Non-Maori patterns were similar although they were more likely than Maori to have come from overseas (less than 5 per cent for both populations), while Maori were more likely than non-Maori to have come from another region.

The context for this mobility is that in the last 30 years of the twentieth century in New Zealand there have been two other very significant transformations. The first has been a Maori cultural revival with associated political and economic changes linked to processes relating to grievances surrounding breaches of the Treaty of Waitangi. The second is the shift from the highly interventionist public policies of the post-war era to the state-led economic restructuring of the 1980s that made New Zealand's economy one of the most vulnerable to external influences anywhere. Both of these transformations had major impacts on Maori and their mobility patterns. Both are integral components of the New Zealand experience of late twentieth century globalization (Kelsey 1995, 1999).

The Maori of the 1970s and the education programmes to ensure survival of *Te Reo Maori* (the Maori language) signalled the clear intention by many Maori to assert their identity. These events must be situated in the wider context of a "revolutionary shift in the development of public discourses and understanding about the nature of this country" (Fleras and Spoonley 1999: x). For Maori perspectives on this transformation see, for example, Awatere (1984), Ihimaera (1995), Melbourne (1995), Walker (1990b), Mead (1997) and Durie (1998).

By the late 1970s reports of young urban-born and raised Maori moving to rural communities to reconnect with their *turangawaewae* and to re-establish links with the *marae* (meeting places) of their ancestral *hapu* illustrated a dimension of the Maori renaissance (Douglas 1979; Stokes 1979; Butterworth 1991). This also showed that Maori were participating in a process of urban–rural migration attracting attention more generally in New Zealand and overseas (Bedford 1983), and showing up in census analysis (Boddington and Khawaja 1993).

There were significant net migration gains in Northland through the 1980s and early 1990s (Pool 1991: 204–7; Scott and Kearns 2000), as well as into the Waikato and Bay of Plenty. This continued into the 1990s producing inflows from inter-regional migration for Northland, Waikato and Bay of Plenty (Te Puni Kokiri 2001).

Migration out of Auckland is closely linked to the impacts that state-led restructuring after 1984 had on New Zealand's urban-based manufacturing industries, and in the 1990s on welfare benefits. Restructuring resulted in the loss of tens of thousands of low-skilled, low-wage jobs for Maori in these industries (Pool and Bedford 1997; Le Heron and Pawson 1996). The effects can be shown to be even more significant when a control is made for the gap between the growth of jobs and increases in the population at working ages (Honey 2001).

Maori incomes had continued to approach non-Maori levels through the 1970s and during a wage-freeze in the early 1980s. But from the mid-1980s until 1996, this trend radically reversed both for all income categories and for incomes of the employed (Martin 1998: 166). Urban Maori were disadvantaged by the restructuring reforms of the 1980s and 1990s, and common responses have been for "discouraged urban workers" either to move out of Auckland into rural areas where there are family/ tribal connections (see Scott and Kearns 2000), or to cross the Tasman to Australia.

International migration of Maori picked up significantly in the late 1970s during an economic recession producing the largest net losses of New Zealanders to overseas destinations on record (Pool and Bedford 1997). The Australian censuses do not have questions on ethnicity on a regular basis, so it is difficult to track changes in Maori numbers in Australia. In 1986, an ethnicity question was included, from which data Lowe (1990: 1) estimated that there were around 26,000 Maori usually resident in Australia, and 75 per cent of them had lived in Australia for less than 10 years (40 per cent for less than 5 years). Employing data from both the 1986 Australian and New Zealand censuses, he also established that "between 1981 and 1986 roughly two Maori migrated from Australia to New Zealand for every three who migrated from New Zealand to Australia". Thus, flows between the two countries have clearly not been uni-directional.

Between 1986 and 2001 the Maori population in Australia is believed to have grown significantly, especially following the loss of over 100,000 secondary-sector jobs in New Zealand between 1986 and 1991. However, there was no Australian census question on ethnicity from 1986 until October 2001. Since the late 1980s it has also not been possible to track trans-Tasman Maori population movement using New Zealand arrival and departure cards, because of the removal of an ethnic self-identity question (Bedford 1987). Circulation of Maori between New Zealand and Australia is thus a significant dimension of their contemporary mobility system but, unfortunately, we cannot document it easily.

Links with Zelinsky

Between 1971 and 2001 the Maori population passed through Zelinsky's Phase C of the "vital" or demographic transition (a major decline in fertility; a continuing but slackening decline in mortality; and a decelerating rate of natural increase). By 2001, Maori society was not yet into Phase D, "the advanced society", when declines in fertility terminate, mortality rates are near fertility levels and natural increase rates are stationary.[13] In the case of mobility, rural–urban migration has slackened, largely because most Maori are now urban residents. International migration is not yet on the decline, as posited in Zelinsky's Phase III, "the late transitional society," but there have definitely been increases in circulation within and between urban areas, rural communities, regions and across the Tasman. Levels of circulation seem to resemble those for the majority society but circular movements appear to be less fluid and complex than the patterns seen, for example, in West Indian societies (Domenach and Picouet 1992) or the tropical Pacific (Bedford 2000).

Although there is no recent assessment of Maori internal migration, comparable to that of Poulson and Johnston (1973) for the late 1960s, it appears from the evidence presented above that "the migratory and circulatory currents of Phase III [of the mobility transition] . . . are much more complex than those of Phase II". Contemporary patterns appear to conform with Zelinsky's (1971: 243–5) scenario:

> the rise of complex migratory and circulatory movements within the urban network, from city to city, or within a single metropolitan region. The gross volume of circulation has undoubtedly risen to a new peak, and many new types of circulation, not all purely economic, have begun to materialise.

Indeed, unlike some modern diasporas the New Zealand pattern (whether Maori or Pakeha) involves flows back and forward for social and cultural as much as for economic reasons.

It is important to recall that Zelinsky (1971) was writing about mobility in "advanced" societies before the latest phase of globalization. Yet, his observations about the third phases of the "vital" and "mobility" transitions still have some relevance for contemporary Maori society. There are some differences in the substance of the migration flows, associated with the Maori cultural renaissance, with the effects of New Zealand's social demography and with New Zealand's special relationship with Australia, but many of the general mobility characteristics Zelinsky identifies for the "late transitional society" apply to Maori. His observations on patterns of population movement during the subsequent two stages – the "advanced" and "super-advanced" societies – are realistic for future Maori mobility as well, although in his 1971 transitional sequence he does not make explicit allowance for a turn-around in migration between urban and rural areas. This return migration to the countryside was not a stream he or other commentators of the day specifically anticipated. Zelinsky points out, however, that he could not predict all future forms of movement for "almost constant change and movement have truly become a way of life".

Paradoxically it could also be argued that Maori society in 2001 has returned to the higher levels of spatial mobility that may have character-ized some of the pre-colonial period. In a sense, they have come a full circle, even if the circuits of movement are now very different, to reach a stage where there is again: "[A] highly complex, intensely interactive social system whose participants . . . indulge in a vast range of irregular, temp-orary excursions, and frequently migrate, in the sense of a formal change of residence" (Zelinsky 1971: 247)

This might seem an outrageous proposition, especially given the com-pletely different demographic, social, economic and political contexts in pre-colonial and post-colonial Aotearoa/New Zealand. Yet, Maori society depended on a wide range of spatial movements for its survival before 1840. It is not entirely irrational to suggest that there may be more simil-arities between the mobility systems in the eighteenth and the late twen-tieth centuries than a model that invokes an "irreversible" series of transi-tional phases might imply. Moreover, Zelinsky's (1979, 1983) revisions acknowledged that "traditional" societies were not necessarily as sedentary as he had originally hypothesized. Complex systems of temporary as well as long-term mobility were common before the onset of "modernization".

That said, three significant differences exist between the phases that characterise the initial and the current stages of the Maori mobility transition:

1   Contemporary Maori society is urban-based, not rural-based, even if cultural affiliations are often with *hapu* that have rural hearths.

2   Maori in 2001 have more migratory options than they had in 1801 within and outside a country they have been obliged to share with a much larger non-Maori population for 140 years.

3   There are four times as many Maori in 2001 as there were in 1801, and all of these people are of mixed ethnic origin (Walker 1990a; Butterworth and Mako 1989). The societies and economies that existed two centuries ago have gone forever, even if the dominant patterns of population movement, especially high levels of circulation, remain common to the mobility systems of both phases.

Mobility transitions and Indigenous peoples: a special case?

Indigenous minorities everywhere have seen their cultures disrupted by foreign intrusions, often associated with imperialism. In the process they have lost control of most of their land, as well as other natural resources forming part of the ecosystems that had sustained them. Following contact, they also suffered significant population changes, most notably increased mortality and numerical declines (Pool 1986). Later, typically after a demographic reprise, many people in these populations, over shorter or longer life-cycle intervals, have moved to urban areas where intra-urban mobility has become the dominant form of population movement.

Given these shared experiences, it is appropriate to posit that there is a variant of the mobility transition common to Indigenous peoples, a hypothesis we would accept albeit admitting that it probably can be sustained only at a generalized level. Our analysis shows that many aspects of Maori mobility would fit such a model, but that their patterns also differ in significant ways from, say, those of Australia's Indigenous peoples. The latter were placed on reserves far from coastal metropolitan centres, where they were seldom a major visible presence and were marginalized, or located in encampments on the peripheries of country towns. Australia had no equivalent of post-war policies attempting to integrate Maori into mainstream New Zealand. Finally, Australia's Indigenous peoples have never migrated internationally at the rates seen for Maori, and Pacific Island Polynesians and Micronesians (Bedford 1992).

It could be postulated that Maori, and perhaps other Indigenous peoples, have undergone transitions shared with all societies subject to colonial intrusion and domination, although each such society's experience will differ as a function of unique circumstances. One could argue, for example, that the Maori transition was closer to that of the Scottish Highlanders after the uprising of 1745, that had pitted Scots and the Crown against Highland "iwi", than to what happened to Indigenous minorities in Anglo-America.

This is not to reject the theoretical power of the mobility transition hypothesis, but it does fit a key point made by Zelinsky (1971: 249) that "the mobility transition, like the vital transition, does not faithfully simulate earlier examples each time it is triggered off". Applied to a particular society, both transitions relate to temporally and spatially bound "episodes in the history of a single large demographic system inter-linked with other social systems, [that] bristle with feedback effects . . . ". Prior events in the transition "make their impact[s] on later phases . . . ", and "earlier or simultaneous events elsewhere[14] can exert a profound effect". Moreover, we would assert, all this is acted out in a particular policy environment.

Many patterns of mobility and vital transition may converge over time; we have argued exactly that here – that the Maori mobility transition is converging towards the Pakeha model.[15] But it is not necessary to assume that societies everywhere will pass through identical sequences. Responding to commentators twenty years after his hypothesis was first articulated, Zelinsky (1993: 219) concluded that "we must accept . . . eccentricities . . . even if we keep chasing after the Holy Grail of grand theory".

Evidence of convergence does appear in the apparent recent upsurges in population circulation among Maori, Pakeha and Pacific Island populations (Bedford 1999, 2000) and further afield such as in the West Indies (Domenach and Picouet 1992), or between Asia and the Gulf States (Skeldon 1997). In one sense new forms of circular mobility resemble traditional patterns of movement, delineated in earlier phases of Zelinsky's model. But there are also totally new elements about them – they are almost "post-modern" in form – in terms of the distances travelled, the means of transport used, and because determinants may include leisure[16] as well as traditional cultural and economic factors. Even in its inter-continental form, this population circulation differs from the great mass migration flows of the turn of the twentieth century or after World War II, notwithstanding the often unacknowledged importance of return migration (Massey 2000). These movements could be seen as manifestations of "individual freedom", guided by the "invisible hand", as posited by neo-liberal public choice theorists.

While not denying the evidence of convergence, and recognizing "post-modern" circulation as a manifestation of this, it also should be clear that mobility is never entirely unconstrained. Among constraints are policy effects,[17] not just direct but also indirect, that are a part of what Zelinsky (1971: 249) called the "social system". Policies typically operate in environments that are likely to be idiosyncratic, an outcome of distinct historical trends in particular social, economic and cultural contexts. It is in part because of these policy effects, direct and indirect, that the Maori mobility transition does not easily fit either the generic Zelinsky model, or some Indigenous peoples' variant.

Notes

1 In contrast Massey *et al.* (1998: vii) barely mention transition theory when "develop[ing] an integrated theoretical understanding of international migration at the end of the twentieth century". There is no reference to demographic or mobility transitions in their subject index, although there is a reference to Zelinsky's (1971) hypothesis in the bibliography to the volume.

2 The international mobility of Aboriginal peoples in Australia, or First Nation peoples in North America does not seem to have as high a profile in the literature.

3 As a singular *iwi* means the population or society as a whole.

4 The attainment of this state has recently been challenged in two interesting papers by Jones and Douglas (1997) and Demeny (1997).

5 Belich (1996: 84) notes that this "neat four-part taxonomy" has been disputed by some recent scholarship.

6 A closely related Polynesian population.

7 The inflows in the late 1860s and 1870s were massive, a result of Premier Sir Julius Vogel's "Public Works" programmes that aimed at "swamping the Maori hinterland" (Belich 1996: 242).

8 One reason New Zealand did not join the Australian Commonwealth (1901) was that from 1840 Maori had all the rights of citizenship, privileges Aboriginal peoples in Australia were not to get until the 1960s (Geddes 1961). Maori men gained the vote in 1867, and this applied virtually to everybody (men, women; Maori, Pakeha) from 1893. Maori parliamentarians can represent either "Maori" or general seats, and Maori opt to vote in one or the other. Maori have frequently held Cabinet rank from the 1890s (NZOYB 1990: 68–9).

9 As noted earlier there may have been a 10 per cent under-enumeration in 1926.

10 Maori were not conscripted for either World War but volunteer rates were extremely high. Maori served both in Maori-led battalions that gained the prestige of Nisei-American units, and across all other services. On the home-front Maori men and women were recruited under wartime labour schemes.

11 Pool (1991: 170–5) makes a similar point about fertility declines, delayed until the 1970s in spite of, at that stage, almost a century of universal education.

12 While the fertility decline was partly related to urbanization, the most powerful determinant was undoubtedly the declines in mortality of 1945–61 (Pool 1991: 170–5).

13 This may be the case for Maori, but the computation of their vital rates is confounded by definitional differences between numerators and denominators (Kukutai 2001).

14 For example, the decision made in Whitehall, under the influence of various factions, to acquire New Zealand as a colony (Belich 1986: 180–7).

15 The same applies in the case of transitions in vital rates (Pool 1991).

16 Circular mobility followed by skiers, surfers and similar groups, continuously across the globe is an example.

17 By policy we mean more than just those direct measures impacting directly on mobility, spanning from the relatively benign policies encouraging Maori post-war migration, to the more prescriptive, acted out each night on the US–Mexico border or the Straits of Gibraltar, through to forced displacement and "ethnic cleansing". In the text we referred to numerous measures indirectly affecting the Maori transition, including as positive factors citizenship, voting, and access to health and social welfare systems, and more negative initiatives such as Vogel's Public Works programme that encouraged mass migration from Europe, and was linked to Land Court processes to wrest land from Maori owners and to allocate it to Pakeha settlers.

References

Awatere, D. (1984) *Maori Sovereignty*, Auckland: Broadsheet Publications.

Bedford, R.D. (1973) 'A transition in circular mobility: population movement in the New Hebrides, 1800–1970', in H.C. Brookfield (ed.) *The Pacific in Transition: Geographical Perspectives on Adaptation and Change*, London: Edward Arnold.

Bedford, R.D. (1983) 'Repopulation of the countryside', in R.D. Bedford and A.P. Sturman (eds) *Canterbury at the Crossroads: Issues for the Eighties*, Christchurch: New Zealand Geographical Society.

Bedford, R.D. (1987) 'Restructuring the arrival card: review and prospect', *New Zealand Population Review*, 13(2): 47–60.

Bedford, R.D. (1992) 'International migration in the South Pacific region', in M.M. Kritz, L.L. Lim and H. Zlotnik (eds) *International Migration Systems: A Global Approach*, Oxford: Clarendon Press.

Bedford, R.D. (1999) 'Culturing territories', in R. Le Heron, L. Murphy, P.C. Forer and M. Goldstone (eds) *Explorations in Human Geography: Encountering Place*, Auckland: Oxford University Press.

Bedford, R.D. (2000) 'Meta-societies, remittance economies and internet addresses: dimensions of contemporary human security in Polynesia', in D.T. Graham and N.K. Poku (eds) *Migration, Globalisation and Human Security*, London: Routledge.

Bedford, R.D. (2001) '2001: reflections on the spatial odysseys of New Zealanders', *New Zealand Geographer*, 57(1): 49–54.

Bedford, R.D. and Heenan, L.D.B. (1987) 'The people of New Zealand: reflections on a revolution', in P.G. Holland and W.B. Johnston (eds) *Southern Approaches: Geography in New Zealand*, Christchurch: New Zealand Geographical Society.

Belich, J. (1986) *The New Zealand Wars and the Victorian Interpretation of Racial Conflict*, Auckland: Auckland University Press.

Belich, J. (1996) *Making Peoples: A History of New Zealanders*, Auckland: Allen Lane.

Belich, J. (2001) *Paradise Reforged*, Auckland: Allen Lane.

Boddington, W. and Khawaja, M. (1993) 'Regional migration during 1986–91, with special emphasis on the Maori population', in D. Brown and C. Mako (eds) *Ethnicity and Gender: Population Trends and Policy Challenges in the 1990s*, Wellington: Population Association of New Zealand and Te Puni Kōkiri.

Brosnan, P. (1986) 'Net inter-provincial migration, 1886–1966', *New Zealand Population Review*, 12(3): 185–204.

Brown, P. (1983) *An Investigation into Official Ethnic Statistics*, Wellington: Department of Statistics.

Butterworth, G.V. (1991) *Nga Take I Neke Ai Te Maori: Maori Mobility*, Wellington: Manatu Maori/Ministry of Maori Affairs.

Butterworth, G.V. and Mako, C. (1989) *Te Hurihanga o Te Ao Maori: Te Ahua o Te Iwi Maori Kuo Whakatautia Maori Development*, Wellington: Department of Maori Affairs.

Chenery, R. and Syrquin, M. (1975) *Patterns of Development, 1950–70*, Oxford: Clarendon Press.

Demeny, P. (1997) 'Replacement-level fertility: the implausible end point of the demographic transition', in G. Jones, R. Douglas, J. Caldwell and R. D'Souza (eds) *The Continuing Demographic Transition*, Oxford: Clarendon Press.

Domenach, H. and Picouet, M. (1992) *La Dimension Migratoire des Antilles*, Paris: Economica.

Douglas, E.M.K. (1979) 'The Maori', *Pacific Viewpoint*, 20(2): 103–10.

Durie, M.H. (1998) *Te Mana Te Kawanatanga: The Politics of Maori Self-Determination*, Auckland: Oxford University Press.

Farmer, R.S.J. (1985) 'International migration', in ESCAP Population Division (eds) *Population of New Zealand*, Country Monograph Series, No. 12, Vol. 1, New York: United Nations.

Fleras, A. and Spoonley, P. (1999) *Recalling Aotearoa. Indigenous Politics and Ethnic Relations in New Zealand*, Auckland: Oxford University Press.

Geddes, W. (1961) 'Maori and Aborigine: a comparison of attitudes and policies', *Australian Journal of Science*, 24(5): 217–25.

Gibson, C. 1973 'Urbanisation in New Zealand', *Demography*, 10(1): 71–84.

Harré, J. (1963) 'The background to race relations in New Zealand', *Race*, 5(July): 3–25.

Harré, J. (1966) *Maori and Pakeha: A Study of Mixed Marriages in New Zealand*, New York: Frederick A Praeger (for the Institute of Race Relations, London).

Hau'ofa, E. (1994) 'Our sea of islands', *The Contemporary Pacific*, 6(1): 147–62.

Hau'ofa, E. (1998) 'The ocean within us', *The Contemporary Pacific*, 10(2): 391–410.

Honey, J. (2001) *New Zealand Jobs, 1976–1996: A Demographic Accounting*, Population Studies Centre Discussion Paper No. 40, Hamilton: University of Waikato.

Hopa, N. (1996) 'The torn *Whariki*', in A.B. Smith and N. Taylor (eds) *Supporting Children and Parents through Family Change*, Dunedin: University of Otago Press.

Howlett, D. (1973) 'Terminal development: from tribalism to peasantry', in H.C. Brookfield (ed.) *The Pacific in Transition: Geographical Perspectives on Adaptation and Change*, London: Edward Arnold.

Hunn, D. (1961) 'Report on Department of Maori Affairs', *Appendices to the Journal of the House of Representatives*, G10, Wellington: Government Printer.

Ihimaera, W. (ed.) (1995) *Vision Aotearoa: Kaupapa New Zealand*, Wellington: Bridget Williams Books.

Jones, G. and Douglas, R. (1997) 'Introduction', in G. Jones, R. Douglas, J. Caldwell, and R. D'Souza (eds) *The Continuing Demographic Transition*, Oxford: Clarendon Press.

Kelsey, J. (1995) *The New Zealand Experiment. A World Model for Structural Adjustment?* Auckland: Auckland University Press and Bridget Williams Books.

Kelsey, J. (1999) *Reclaiming the Future: New Zealand and the Global Economy*, Wellington: Bridget Williams Books.

Kukutai, T. (2001) 'Maori identity and "Political Arithmetick": The dynamics of reporting ethnicity', MSocSc Thesis in Demography, University of Waikato.

Le Heron, R. and Pawson, E. (1996) *Changing Places. New Zealand in the Nineties*, Auckland: Longman Paul.

Lewis, D. (1972) *We, the Navigators*, Auckland: A.H. and A.W. Reed.

Lowe, J. (1990) *The Australian Maori Population: A Demographic Analysis Based on 1986 Australian and New Zealand Census Data*, Report for the Population Monitoring Group, Wellington: New Zealand Planning Council.

McCreary, J. (1972) 'Population growth and urbanization', in E. Schwimmer (ed.) *The Maori People in the 1960s*, Auckland: Longman Paul.

Marsault, M., Pool, I., Dharmalingham, A., Hillcoat-Nalletamby, S., Johnstone, K., Smith, C. and George, M. (1997) *Technical and Methodological Report*, Population Studies Centre Technical Report Series No. 1, Hamilton: University of Waikato.

Martin, B. (1998) 'Incomes of individuals and families in New Zealand, 1951–96', unpublished PhD Thesis in Demography, University of Waikato.

Massey, D.S. (2000) 'Immigration and globalization: policies for a new century', invited paper presented at the *Migration: Scenarios for the 21st Century International Conference*, Rome, 12–14 July 2000.

Massey, D.S., Arango, J., Hugo, G., Kouaouci, A., Pellegrino, A. and Taylor, J.E.

(1998) *Worlds in Motion: Understanding International Migration at the End of the Millennium*, Oxford: Clarendon Press.

Mead, S.M. (1997) *Landmarks, Bridges and Visions: Aspects of Maori Culture*, Wellington: Victoria University Press.

Melbourne, H. (1995) *Maori Sovereignty: Maori Perspectives*, Auckland: Hodder Moa Beckett.

Metge, A.J. (1952) 'The Maori population of northern New Zealand', *New Zealand Geographer* 8(2): 104–24.

Metge, A.J. (1964) *A New Maori Migration: Rural and Urban Relations in Northern New Zealand*, London: Athlone Press and Melbourne University Press.

Metge, A.J. (1971) *The Maoris of New Zealand*, London: Routledge and Kegan Paul.

Metge, A.J. (1995) *New Growth from Old: The Whanau in the Modern World*, Wellington: Victoria University Press.

NZOYB (1990) *New Zealand Official Year Book*, Wellington: Department of Statistics.

Notestein, F.W. (1945) 'Population: the long view', in T.W. Schultz (ed.) *Food for the World*, Chicago: Chicago University Press.

Omran, A.R. (1982) 'Epidemiologic transition: theory', in J.A. Ross (ed.) *International Encyclopaedia of Population*, Vol. 1, New York: Free Press.

Pearson, W. (1962) 'New Zealand since the war: 7. The Maori people', *Landfall*, 16(2 June): 148–80.

Petersen, W. (1961) *Population*, New York: Macmillan.

Pool, I. (1961) 'Maoris in Auckland', *Journal of the Polynesian Society*, 70(1): 43–66.

Pool, I. (1963) 'When is a Maori a 'Maori' A viewpoint on the definition of the word Maori', *Journal of the Polynesian Society*, 72(3): 206–10

Pool, I. (1977) *The Maori Population of New Zealand, 1769–1971*, Auckland: Auckland University Press.

Pool, I. (1986) 'The demography of indigenous minority populations', *Proceedings of General Conference, International Union for the Scientific Study of Population, Florence, 1985*, 4 vols, Liege: IUSSP, 1: 135–41.

Pool, I. (1991) *Te Iwi Maori*, Auckland: Auckland University Press.

Pool, I. (2000a) 'Vers un modele de la 'Transition-Age Structurelle': une conse-quence mais aussi une composante de la Transition Demographique', Conference invite, l'Association Canadienne-francaise pour l'Avancement de Science, Montreal, May 2000 (under editorial review).

Pool, I. (2000b) 'Age-structural transitions and policy: frameworks', paper presented at the Seminar IUSSP Committee on Age-structures and Policy and Asian Population Network, Phuket, November 2000.

Pool, I. and Bedford, R.D. (1997) 'Population change and the role of immigration', in *Proceedings of the Population Conference, 12–14 November, Wellington*, Wellington: New Zealand Immigration Service, 62–117.

Pool, I., Dickson, J., Dharmalingam, A., Hillcoat-Nalletamby, S., Johnstone, K. and Roberts, H. (1999) *New Zealand's Contraceptive Revolution*, Hamilton: Social Science Monograph Series, Population Studies.

Poot, J. (1984) 'Models of New Zealand internal migration and residential mobility', unpublished PhD Thesis in Economics, Victoria University of Wellington.

Poulson, M. and Johnston, R.J. (1973) 'Patterns of Maori migration', in R.J. Johnston (ed.) *Urbanisation in New Zealand*, Wellington: Reed Education.

Poulson, M., Rowland, D.T. and Johnston, R.J. (1975) 'Patterns of Maori migration in New Zealand', in L.A. Kosinski and R.M. Prothero (eds) *People on the Move*, London: Methuen.

Rowland, D.T. (1971a) 'The age structures of Maoris in Auckland', *Proceedings of the Sixth New Zealand Geography Conference*, Christchurch: New Zealand Geographical Society.

Rowland, D.T. (1971b) 'Maori Migration to Auckland', *New Zealand Geographer*, 27(1): 21–37.

Rowland, D.T. (1972) 'Processes of Maori urbanization', *New Zealand Geographer*, 28(1): 1–22.

Rowland, D.T. (1973) 'Maori and Pacific Islanders in Auckland', in R.J. Johnston (ed.) *Urbanization in New Zealand: Geographical Essays*, Wellington: Reed Education.

Rowland, D.T. (1974) 'The family status of Maori in Auckland', *Australian Geographical Studies*, 12(1): 27–37.

Scott, K. and Kearns, R. (2000) 'Coming home: return migration by Maori to the Mangakahia Valley, Northland', *New Zealand Population Review*, 26(2): 21–44.

Skeldon, R. (1985) 'Population pressure, migration and socio-economic change in mountainous environments: regions of refuge in comparative perspective', *Mountain Research and Development*, 5(3): 233–50.

Skeldon, R. (1990) *Population Mobility in Developing Countries: A Reinterpretation*, Bellhaven: London.

Skeldon, R. (1992) 'On mobility and fertility transitions in East and Southeast Asia', *Asian and Pacific Migration Journal*, 1(2): 220–49.

Skeldon, R. (1997) *Migration and Development: A Global Perspective*, Harlow: Addison Wesley Longman.

Sorrenson, M.P.K. (1956) 'Land purchase methods and the effects on Maori population, 1865–1901', *Journal of the Polynesian Society*, 65(3): 183–99.

Stokes, E. (1979) 'Population change and rural development: the Maori situation', in R.D. Bedford (ed.) *New Zealand Rural Society in the 1970s: Studies in Rural Change No. 1*, Christchurch: Department of Geography, University of Canterbury.

Stone, R.C.J. (2001) *From Tamaki-Makau-Rau to Auckland*, Auckland: Auckland University Press.

Taylor, J. and Bell, M. (1996) 'Population mobility and indigenous people: the view from Australia', *International Journal of Population Geography* 2(2): 153–70.

Te Puni Kokiri (2001) *Te Maori I Nga Rohe. Maori Regional Diversity*, Wellington: Ministry of Maori Affairs.

Te Rangihiroa [Sir Peter Buck] (1952) *The Coming of the Maori*, Wellington: Government Printer.

Thompson, B. (1985) 'Industrial structure of the labour force', in ESCAP Population Division (eds) *Population of New Zealand*, Country Monograph Series, No. 12, Vol. 2, New York: United Nations.

Thompson, R.H.T. (1963) *Race Relations in New Zealand: A Review of the Literature*, Christchurch: National Council of Churches.

Urlich, D. (1969) 'Maori population and migration, 1800–40', MA Thesis in Geography, University of Auckland.

Vayda, P. (1960) *Maori Warfare*, Wellington: Polynesian Society.

Waitangi Tribunal (1997) *Muriwhenua Land Report*, Wellington: Waitangi Tribunal.

Walker, R. (1990a) 'Cultural Continuities', *NZ Listener*, April 16: 96–97.

Walker, R. (1990b) *Ka Whawhai Tonu Matoa: Struggle Without End*, Auckland: Penguin Books.
Walker, R. (1996) *Nga Pepa a Ranginui: The Walker Papers*, Auckland: Penguin Books.
Walker, R. (2002) In the *NZ Herald*, 6 February.
Watson, M. (1985) 'Urbanization', in ESCAP Population Division (eds) *Population of New Zealand*, Country Monograph Series, No. 12, Vol. 1, New York: United Nations.
Woods, N. (1960) 'Immigration and the labor force', *Industrial Development Conference*, Background Paper No. 26, Wellington: Department of Industries and Commerce.
Wright, H. (1959) *New Zealand 1769–1840: The Early Years of Contact*, Cambridge, Mass: Harvard University Press.
Zelinsky, W. (1971) 'The hypothesis of the mobility transition', *Geographical Review*, 61(2): 219–49.
Zelinsky, W. (1979) 'The demographic transition: changing patterns of migration', in P.A. Morrison (ed.) *Population Science in the Service of Mankind*, Liege: International Union for the Scientific Study of Population.
Zelinsky, W. (1983) 'The impasse in migration theory: a sketch map for potential escapees', in P.A. Morrison (ed.) *Population Movements: Their Forms and Functions in Urbanization and Development*, Brussels: Ordina Editions.
Zelinsky, W. (1993) 'Classics in human geography revisited – author's response', *Progress in Human Geography*, 17: 217–19.

4 Migration and spatial distribution of American Indians in the twentieth century

Karl Eschbach

The spatial distribution and migration behaviour of American Indians has long occupied an important place in the study of American Indian experiences in the twentieth-century United States. American Indians began the century with a distinctive spatial distribution compared to other American racial populations. In 1900, the majority of American Indians lived in rural areas west of the Mississippi River, and with a high degree of geographical isolation from other Americans. By the end of the century, the Indian population had substantially redistributed, though even at the end of the century a large segment of the Indian population continued to live in rural areas in states between the Mississippi and the Pacific coast of the United States.

The context of twentieth-century American Indian migration

The spatial distribution and migration of Indians has been an issue of central importance as a legacy of the nineteenth century policies of Indian Removal and the establishment of the reservation. These two policies were adopted by the United States government to resolve competition for control of the American continent between Native Americans and European American interlopers. The removal policy was adopted in the Indian Removal Act of 1830, which mandated the forced removal of Indians from the southeastern United States east of the Mississippi River. Indians forced to migrate under terms of the Removal Act were primarily resettled in Indian Territory, which later formed the majority of the state of Oklahoma. While Indian removal was intended to permanently remove Indian tribes from beyond the zone of intensive Euro-American settlement, westward expansion of this settlement in the 1840s quickly rendered the solution obsolete. The reservation system arose in response. Under this system, Indian tribes were resettled on designated lands, either within their former territorial range or, sometimes, in a different location. In exchange tribes were given promises of territorial sovereignty and economic assistance (Prucha 1984; White 1991).

The impact of Indian removal and the reservation system on Indian populations has been manifold. First, by isolating Indian populations geographically, these policies raised spatial barriers to the integration of Indians into the mainstream economic and social life of the United States, despite the professed intentions and policies of the United States Indian Office (later Bureau of Indian Affairs) to the contrary. Second, the bargains struck between the United States and Indian tribes to implement the removal and reservation policies gave Indian tribes a legal claim to a limited territorial sovereignty in the lands in Indian Territory and on reservations that were set aside for them. This fact would have important implications at the end of the twentieth century, when Indians were increasingly able to hold the United States to its century-old promises of sovereignty. Third, the institutionalization of a system of legal recognition of Indian tribes that were sovereign in reservation land bases created a geographical dimension to the recognition of American Indian identity. Each of these three points has been the centre of persistent tensions in twentieth-century American Indian policy, and more broadly in the experiences of Indians in American society.

Academic study about American Indian migration has focused on the nexus between reservation communities and urban areas. Issues such as the timing of urbanizing migration flows, the relationship between migration patterns and reservation social and economic conditions, the selectivity of urban migration, the purposes and life time patterns of migration, the effect of migration on the individual migrant, and the situation of urban Indians have all received considerable attention. The literature on American Indian migration peaked in the period between the 1950s and 1970s, coincidentally with the rapid urbanization of the American Indian population. More recently, studies of Indian migration have had less focus, as American Indian migration patterns have become more complex and less easy to understand within the "reservation-to-city" framework.

The pendulum of United States Indian policy: sovereignty or assimilation?

The system of removal to reservations in the nineteenth century represented a pragmatic, and largely incoherent, compromise between policy goals: first, removing the impediment of native occupancy and title to the expansion of Euro-American settlement across the American continent, and second, assimilating Indians into American civil society. Indian removal and the reservation accomplished the first goal expediently, and by the end of the 1880s Indians ceased to be a military threat to the manifest destiny of Euro-American expansion (White 1991). However, by this time, the primary goal of United States Indian policy had become the full assimilation of Indians into American society *as individuals*, despite promises of continued tribal sovereignty made during the process of treaty-making that

had preceded reservation resettlement (Hoxie 1984). The institutionalization of tribal sovereignty, and the social, institutional, and geographical
isolation of reservation populations served rather to frustrate than to
accomplish this goal. Tension between the promises of tribal sovereignty in
reservation places and the incorporation of Indians into American society
remained a ubiquitous backdrop to reservation social conditions throughout the twentieth century.

One consequence of this tension was a perpetual pendulum swing in
United States Indian policy between attempts to disrupt and to empower
Indian tribal government. The 1887 Dawes Act created the framework for
assimilationist policies that held sway until the 1920s. Chief among these
policies mandated by the Dawes Act was the "allotment" of reservation lands
held in trust for tribes to individual Indians in order to encourage them to
farm fee-simple freeholds using Euro-American agricultural methods. Lands
declared to be surplus after allotment were to be sold to non-Indians. The
allotment policy was accompanied by paternalistic attempts by United States
authorities to disrupt Indigenous social and cultural organization and to
acculturate individual Indians. By 1934, the allotment policy had led to the
alienation of more than 80 million acres of Indian lands (well over half the
1880 Indian land base), but not the widespread creation of an agricultural (or
any other) economic base to support reservation populations (McDonnell
1991).

The passage of the Indian Reorganization Act during the "Indian New
Deal" in the 1930s marked a shift in policy. This policy encouraged Indian
tribes to reorganize tribal governments on western elective models, and to
form business corporations to manage tribal assets. These policies marked
an important symbolic shift, because they recognized the collective organization of Indian societies as an asset rather than an impediment to the
incorporation of Indians into American society. Concomitantly, in-place
collective development of reservation resources was recognized in principle
as a component of the solution to the chronic material poverty of reservation populations (Cornell 1988; Prucha 1984).

By the late 1940s, another pendulum shift led to a renewed emphasis in
United States Indian policy on the *termination* of federal trust responsibility
to Indian tribes. Tribes that appeared to be relatively successful in developing an economic base were encouraged to graduate from trust status,
ending specific federal jurisdiction and responsibility to the designated
tribal populations. The United States did terminate recognition of several
tribes in the 1950s (Tyler 1973). Beginning in the late 1940s, the federal
government also began to subsidize migration of Indians from reservation
communities to urban areas. This policy sought to capitalize on the perception that Indians who had served in the armed forces in World War II
were receptive to off-reservation employment opportunities, and reflected
the belief that out-migration from reservation to expanding urban labour
markets was a practical remedy to reservation underdevelopment (Tyler

1973). Between 1952 and 1972, the Bureau of Indian Affairs' relocation and employment assistance programmes had assisted more than 100,000 Indians in their move from reservations to urban labour markets, through both direct subsidy of relocation expenses and by the provision of job training in industrial skills valued by urban employers (Sorkin 1978: 25).

The termination and relocation policies themselves were progressively undermined by several concordant changes in post-World War II American society. One important shift was the growth of social service provision by the federal government. Even during the 1950s, the terminationist agenda of the Eisenhower Bureau of Indian Affairs was undercut by the sheer growth of federal spending for health and welfare, which increased the flow of resources to reservation populations (Nagel 1996). In addition, the paternalism that had been one of the chief attributes of federal Indian policy since the establishment of the reservation system was increasingly out of step with the civil rights revolution of the 1950s and 1960s. Tribes brought pressure on the United States government to empower their own leaders to control Indian policy, in place of appointed non-Indian bureaucrats. This pressure led to the emergence of a policy of tribal "self-determination without termination" which was substantially mandated by the Indian Self-determination and Education Act of 1976 (Castile 1998; Nagel 1996).

Commentators have questioned the depth of federal commitment to this principle. Verbal recognition of tribal self-determination has been accompanied at times by diminishing federal appropriations to support programmes for Indians, the narrowing of tribal jurisdiction, and by increasing pressure to limit the membership and recognition of Indian tribes (Barsh 1994). Nonetheless, recognition of the principle of tribal self-determination rendered direct termination and relocation policies increasingly irrelevant to federal Indian policy, because neither policy was central to the concerns of tribal leaders.

Regional redistribution

From the establishment of the reservation system through the census of 1940, the Indian populations remained rooted in rural communities located on or near to designated reservations, and in areas of historic tribal jurisdiction in Oklahoma. Precise measurements of migration flows away from reservation areas in this period are not available. The United States census does not contain an item to measure migration until 1950. However, as late as 1940, the census enumerated just 8 per cent of Indian populations in urban areas, implying extremely low out-migration totals from rural reservation areas. By contrast, by 1940 nearly three-fifths of the White and one-half of the African American population lived in urban areas (Shoemaker 1999: 77). This small number of urban Indians to some extent reflects an under-enumeration of Indians in off-reservation urban settings (an issue discussed in more detail below). By any reasonable inference, however, urbanizing out-migration from reservation communities was minimal in this period when compared to the rapid pace of urbanizing of other American populations.

After 1940, Indian out-migration from reservation areas increased dramatically, for several reasons. First, participation by American Indians in military service and in war industries during World War II, dramatically increased the exposure of native peoples to non-Indian society, and consequently increased the comfort of Indians in social relationships with non-Indians, raising aspirations for material standard of living, and increasing demands for equitable treatment from the mainstream institutions of American civil society. Second, the progressive narrowing of socio-cultural differences between Indians and non-Indians in the post-war period steadily lowered the barriers to Indian participation in social and economic relationships with non-Indians. Third, and perhaps least importantly in quantitative terms, the Bureau of Indian Affairs Indian relocation programme subsidized this relocation (Bernstein 1991; Fixico 1986).

After 1940, the spatial distribution of the American Indian population changed abruptly. Between 1940 and 1980, the urban percentage of the American Indian population increased from 8 per cent to 53 per cent. This latter figure was still well shy of the 74 per cent of the United States population that lived in urban areas, but indicates an extremely rapid urbanization of the Indian population. The percentage of Indians living in metropolitan areas increased from 39 per cent in 1970 to 51 per cent in 1990, and then to 66 per cent in 2000 (Shoemaker 1999).

The regional distribution of the American Indian population also shifted significantly after 1940. Table 4.1 shows the distribution of American Indians according to each of nine census divisions (groupings of American states used by the Census Bureau to report regional population distributions). The three census divisions that lie to the west of the Mississippi River encompass the majority of reservation lands and are indicated in Table 4.1 as "Central Reservation divisions". The Indian population of the states in these divisions declined from 70 per cent of the total Indian population to 52 per cent in 1990, and 43 per cent in 2000. The five census divisions that lie to the east of the Mississippi River include many states from which Indian populations had been removed under the terms of the Indian Removal Act of 1830. These states contained just 17 per cent of the Indian population in 1940, 25 per cent of the Indian population in 1990, and 32 per cent in 2000. The Indian population in the five Pacific Rim states increased from 13 per cent of the total in 1940 to 23 per cent in 1990, and 25 per cent in 2000.

Migration and changes in identity

Changes in spatial distribution of the American Indian population in the second half of the twentieth century have frequently been attributed to

Table 4.1 American Indian population by census division: 1940, 1990 and 2000

Census division	1940	1990	2000		
			All	Single race reported	Multiple races reported
West North Central	15.0	9.6	8.1	9.0	6.7
West South Central	19.2	17.8	16.7	17.6	15.4
Mountain	35.3	24.5	18.3	24.8	8.5
Central 'reservation' divisions	69.5	52.0	43.1	51.4	30.6
New England	0.7	1.7	2.4	1.7	3.6
Middle Atlantic	2.7	4.7	6.6	4.9	9.3
South Atlantic	7.3	8.8	10.8	9.4	12.8
East South Central	0.8	2.1	3.1	2.3	4.3
East North Central	5.7	7.7	9.3	7.1	12.5
Eastern 'removal' divisions	17.2	24.9	32.2	25.5	42.4
Pacific[a]	13.3	23.1	24.7	23.2	26.9
Total per cent	100.0	100.0	100.0	100.0	100.0
Number of Indians	345,252	1,959,234	4,119,301	2,475,956	1,643,345

Sources: Thornton (1987); US Bureau of the Census (1992, 2001).

Note:

[a] 1940 data include the Indian population of Alaska and Hawaii, which were included in the Pacific Census Division after achieving statehood in 1859.

migration flows. This is particularly the case with respect to the urbaniz-ation series, which has often been reported as a straightforward indicator of the volume of urbanizing migration from reservations to cities. For example, Sorkin (1978: 1) began a monograph on urban American Indians with the observation that "most urban Indians were born or raised on reservations and subsequently moved to the city". However, the emergence of an urban, off-reservation Indian population in the post-World War II period was a more complex process that occurred both because of urbanizing migration from reservations, but also because of rapid changes in the social and cultural climate for the recognition of American Indian identity in urban areas, and in regions of the country from which Indians had once ostensibly been removed.

One of the most important, but little remarked, implications of the Indian removal and reservation system was the constriction of recognition of American Indian identity to those places to which tribal populations had been relocated. The institutionalization of Indian tribes as quasi-sovereign "domestic dependent nations" gave tribes a corporate identity that was unique among American racial and ethnic populations. The obverse impli-cation of this system was that Indian individuals who did not participate in one of the recognized tribes had an uncertain status as Indians. The United States government was eager to overlook the persistence of tribal groups that it had not formally recognized as part of the reservation system, and to declare that Indian individuals who left reservations to live and work among European Americans had "broken tribal relations", thereby ending any personal claims as beneficiaries of the trust relationship between the United States government and the recognized tribes.

Indian removal and the reservation system overlooked many Indian individuals. In the southeastern United States, in particular, removal left behind many Indian individuals in a variety of circumstances ranging from ecologically distinct rural hamlets to residential propinquity and amalgamation with European Americans and African Americans. Ethnic relations between Indians and non-Indians in off-reservation settings have long been relatively fluid, with very high rates of intermarriage with other racial populations (Eschbach 1995; Thornton 1987). This led to the rapid growth of a mixed ancestry off-reservation Indian population that was much larger than the officially recognized reservation-resident Indian population that was the object of federal Indian policy. Through the first half of the twentieth century, the Indian identity of these off-reservation Indians was overlooked by a bureaucratic system that was focused on reservation populations.

The burgeoning out-migration of Indians from reservations after 1940 coincided with increasing institutional and cultural recognition of off-reservation populations. In many respects, the changes amounted to the adoption by Indian individuals of the kind of voluntary and private ethnic pluralism that had long been expressed by European immigrant populations. The ethnic bargain that underlay the melding of European immigrants was to permit and even encourage individuals to proclaim personal ethnic loyalties, though with few consequences for the relationship of the individual with government authority (Fuchs 1990; Waters 1990). For Indians, by contrast, the reservation system had implied a more formal, bureaucratically managed concept of Indian identity that required participation in recognized tribes, and that ignored subjective personal identity. While this system of bureaucratic recognition persists, changing cultural conceptions of ethnic identity and growing minority ethnic pride encouraged the expression of ethnic identification as an Indian among off-reservation mixed ancestry Indians (Nagel 1996).

In 1960, the mode of administration of the United States census changed from enumerator coding in a face-to-face interview to respondent self-report or a mail-out, mail-back form. This change converted the census race item from an indicator of enumerator perceptions to a question about subjective personal identity. This change occurred in the context of evolving perceptions of the meaning of American Indian identity that encouraged increased identification with Indian identity among off-reservation Indians. Subsequent changes in the racial classification of persons with mixed Indian and non-Indian ancestry were among the most important sources of spatial redistribution of the Indian population. Between 1960 and 1990, the American Indian population increased from 345,000 to 1.96 million. Passel (1996) estimated that almost half of this increase (645,000 persons) was an "error of closure" that implied increased self-identification as an Indian rather than the excess of new births over deaths (see also Passel and Burman 1986). Then for the 2000 census, the Census Bureau permitted the reporting of two or more racial identities for the first time. The Indian population ballooned to 4.12 million, with 1.64 million persons reporting American Indian identity along with one or more additional racial identities (US Bureau of the Census 2001).

The importance of these changes in self-identity for data on spatial distribution of the Indian population is that the increases tended to occur primarily outside of reservation areas as shown by comparing the distributions in Figures 4.1 and 4.2. A substantial portion of the increases occurred in those eastern – particularly southeastern – states from which Indians had ostensibly been removed a century and a half earlier (Eschbach 1993; Passel 1996). Multiple race reporting by Indians in 2000 was most popular in Eastern states, and least popular in states with large reservation populations. Multiple race reporting was also much more popular in metropolitan areas (47 per cent of Indians reported two or more races) than in non-metropolitan areas (26 per cent report two or more races). Just 15 per cent of reservation Indians reported two or more races, compared to 47 per

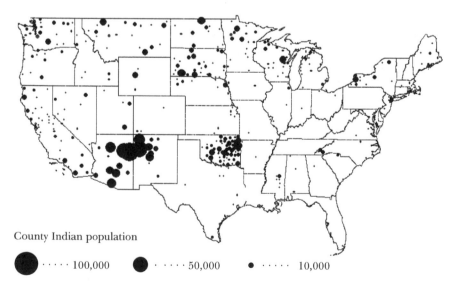

County Indian population

● ····· 100,000　　● ····· 50,000　　• ····· 10,000

Figure 4.1　County population of American Indians in the continental United
　　　　　　States, 1950. Symbols are scaled by the square root of the county
　　　　　　population of American Indians.

Source: US Bureau of the Census, 1950 Census of Population.

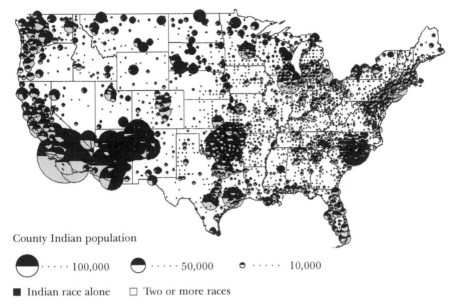

County Indian population

● ·····100,000   ● ·····50,000   ○ ····· 10,000

■ Indian race alone   □ Two or more races

Figure 4.2 County population of American Indians in the continental United
States, 2000. Symbols are scaled by the square root of the county
population of American Indians alone or in combination with some
other race.

Source: US Bureau of the Census, 2000 Census of Population Summary File 1.

cent of Indians who lived outside of reservations (US Bureau of the Census
2001).

The substantive contribution of changes in identification to population
redistribution is clearest with respect to changes in regional distribution by
census divisions. Eschbach (1993) reported an analysis of lifetime migration
(the contrast between state of birth and state of residence) by Indians using
1980 census data. Taken at face value, the data show that eastern states
were by a small margin net out-migration areas for American Indians in the
population that was alive in 1980, despite the sharp growth of population
share in these states in the preceding decades. There was substantial in-
migration to California and other western states. Approximately one-third
of older Indians (born before 1930) living in western states came from
states with large reservation populations. Approximately one-quarter were
born in eastern areas without large reservation populations. By 1980, a
majority of Indians living in urban centres in California and Illinois were
born in their current state of residence (US Bureau of the Census 1993).
These data imply that all of the increased Indian population share of
eastern states is attributable to changes in ethnic classification rather than
to net in-migration, and that new migration from reservation areas contri-

buted a minority of Indian population growth in cities in California and elsewhere on the Pacific coast.

Less definitive statements are warranted about the sources of urban Indian population growth than about regional shifts in the Indian population. This is the case because there is little information from the census, or from other sources, about Indian population movements within states. The best available information suggests that urbanizing flows from reservation to city did increase dramatically after 1940, forming an important component of the growth of urban Indian populations.

The most detailed study of spatial distribution and migration for a single Indian tribe used a population register for the Tohono O'odham of southern Arizona that was assembled in the 1960s. This register was intended to include a record of demographic events for all descendents of nineteenth century tribal rolls. The study found that 49 per cent of all Indians who were born in a reservation village permanently resettled outside of their village of birth. Permanent resettlement increased from 27 per cent of Indians born before 1900, to 54 per cent of Indians born during the decade of the 1930s. A third of the Indians in the register removed to an off-reservation setting, with sharply rising rates of off-reservation migration in later cohorts. Three conclusions are warranted from these findings. First, there was a sizeable volume of out-migration from the reservation, even in the earliest decades of the twentieth century. Second, the period around 1940 did seem to mark a watershed in the increase in out-migration from reservation communities. Hackenberg reported that the stimulus to increased permanent out-migration by the Tohono O'odham was the growing availability of non-seasonal jobs outside of the reservation. Third, out-migration rates from Tohono O'odham lands remained relatively low compared to other rural Americans, even in the 1940s and 1950s, and especially given the chronic unemployment and material poverty experienced by reservation residents (Hackenberg 1972). Unfortunately, there has been little comparative research that has been equally rigorous to discover whether Tohono O'odham patterns were general to other reservation populations. There has been no subsequent research to determine how migration patterns have changed since the 1960s.

Understanding the relative contribution of migration and changes in identification to Indian population redistribution in the post-war period is crucial to our understanding of both the emerging urban Indian populations in the post-war period, and the evolving demography of reservation communities. Contemporaneous scholarly literature about the post-war urbanization of the Indian population probably over-emphasized new migration from reservations as the source of urban Indian population growth. Concomitantly, it probably over-emphasized the abruptness of the transition to urban life experienced by the urban Indians by over-generalizing from the experiences of that segment of this population that were new urban arrivals from socially and culturally isolated reservation communities.

Finally, over-emphasis on reservation out-migration as the sole component of rapidly increasing Indian population implies out-migration rates from reservations that were unrealistically high. Hackenberg and associate's study of the Tohono O'odham showed that moderate out-migration rates from the tribe's reservation of the level observed for the 1940s and 1950s were consistent with continued growth of the reservation's population, because of the relatively youthful age structure and high levels of fertility of that population. This finding does appear to generalize to other reservation populations, which continued to grow even during the high point of Indian urbanizing migration (Shoemaker 1999).

Migration from reservations

Interest in American Indian migration peaked in the period from the end of World War II to 1980. Several concurrent causes created this interest. First, as already noted, out-migration from reservation communities increased markedly in the aftermath of World War II. Second, the Indian Relocation Program marked the first time that the federal government actively worked on a broad scale to diminish Indian concentration on reservations by subsidizing out-migration on a broad scale. A portion of the research on American Indian migration was conducted to evaluate the successes and failures of the relocation programme. Third, increasing attention to minority rights issues raised consciousness of the condition of American Indians in all settings. Many of the voices that contributed to Indian rights movements in the period came out of urban settings and highlighted a broader set of concerns than the conditions of reservation Indians.

The primary focus of empirical studies of American Indian migration was the nexus between reservations and similar rural Indian enclaves and urban labour markets. Scholars addressed issues of the determinants of migration decisions, the impact of migration and non-migration on social and economic conditions on reservations, the migratory strategies pursued by Indian populations, and the process of adjustment of Indians who were newly introduced into off-reservation settings.

One important theme in the study of migration to and from reservations has been the economic impact of low rates of out-migration on reservation communities. The development of an economic and employment base for reservations suitable to sustain material standards of living comparable to other Americans has long been seen to be a precondition for reservations to be successfully self-governing communities (Kalt and Cornell 1992). Out-migration to reduce excess labour supply in labour markets is typically a component of non-metropolitan development, and the reluctance of Indians to migrate from reservations is thus a contributing explanation of reservation underdevelopment. The issue is controversial because urbanizing out-migration as a development strategy seems to compromise the promise of tribal sovereignty. For this reason, the Indian Relocation Program and similar pro-migration policies have sometimes been interpreted as a terminationist alternative to subsidized in-place development (Fixico 1986; Tyler 1973). These issues are likely to remain of considerable importance to American Indian policy debates given the continued recognition of the sovereignty of reservation-based governments, and the continued under-development of reservation economies.

The empirical basis and theoretical terms of this debate are uncertain, for several reasons. First, in-place development and out-migration from rural communities are more likely to be concomitants than mutually exclusive strategies – successful rural development may increase rather than substitute for out-migration by increasing the integration of rural and urban labour markets. Second, it is not clear that urbanizing migration is an economically optimal response for Indian communities and individuals. Using 1980 census data, Snipp and Sandefur (1988) found no economic payoff to out-migration from reservation communities, at least within the first five years of migration. Similarly, it is difficult to establish a correlation between reservation out-migration rates and unemployment rates, because reservation out-migration rates may not (only) be adaptive responses to economic conditions that equilibrate labour markets.

The majority of empirical studies of American Indian migration were written between the 1950s and 1970s. Waddell and Watson (1971) contains a representative selection of this literature. Useful summaries of the findings of this literature include Thornton *et al.* (1982), and a recent monograph on urban American Indians by Fixico (2000). The hallmark of this literature was an emphasis on the transitional experiences of urban American Indians. What was striking about this population was that it was new. Urban Indians were seen primarily as an *urbanizing* population coming out of isolated rural environments where they had had little exposure to the cultural practices of urban, non-Indian America. A segment of this literature was written to analyse and evaluate the success and failure of the Indian Relocation Program (Ablon 1964; Neils 1971; Sorkin 1978). One important question raised in the literature included the migration strategies used by the migrants, that is, whether migration was intended to be circular or permanent. A related set of issues concerned the selectivity and outcome of migration. Who migrated? Who returned? What was the rate of persistence in urban environments or of return to reservation communities? Another important set of concerns was the quality of the adjustment to urban environments, as reflected, for example, in economic attainment, housing quality, rates of arrest, alcoholism and mental health morbidity. Another theme in this literature examined the institutional and personal strategies used to adapt to urban conditions. What was the role of social institutions such as Indian centres, powwow clubs, and churches in maintaining and transforming the Indian identity. What were the informal patterns of response in terms of settlement in urban enclaves, participation in ethnically dense social networks? Finally, what were the consequences of urban conditions and adaptive responses for personal ethnic identity, for the formation of inter-tribal relationships and identity, and for social and cultural participation in the urbanizing migrants' own tribes in the communities left behind.

Urban Indians as migrated Indians

In the 1928 Brooking Institution report of the social and economic conditions of reservation Indians, the chapter on urban Indians was entitled "Migrated Indians". This usage is a peculiar trope that is a byproduct of the reservations system. The reservation system created a distinct geography of nations-within-a-nation where the tribal sovereignty has its strongest reach. Many Indians never participated in this geography because the United States government did not recognize them as tribes when the reservation system was created, or because as individuals their lives were already disconnected from the recognized tribal groups. Many other Indians were drawn away from reservations because the geography of tribal sovereignty was so sharply disengaged from the emergent twentieth century geography of economic opportunity. Indians who live in urban and metropolitan centres away from the recognized tribal lands are in a sense symbolically "migrated Indians" because they are out of place with respect to the bureaucratic geography of tribal recognition. The relationship of these migrated Indian to organized tribes and to the government of the United States remains a persistent and unsolved question.

Off-reservation, urban Indians are now a distinct majority of the American Indian population. In 1990, the census Indian population living in metropolitan areas comprised 51 per cent of the national Indian population of 1.96 million. In 2000, 2.7 million persons reporting an Indian race alone or in combination lived in metropolitan areas. This was two-thirds of the 4.12 million American Indian population. Fifty-seven per cent of persons who reported American Indian race alone – 1.42 million people – lived in a metropolitan area. One-quarter of the Indian population – 1.03 million people – were enumerated on reservations or other Indian trust lands or areas of historic tribal jurisdiction. This was an increase from 440,000 reported in 1990 (US Bureau of the Census 1993), though much of the increase can be attributed to the Census Bureau's increased thoroughness in identifying Indian lands.

Fully 78 per cent of the multiple-race Indian population identified by the 2000 census lived in metropolitan areas, and there is strong indication that new reporting of Indian identity by mixed ancestry Indians who were already living in cities, rather than new migration to urban areas, will explain a large portion of this apparent increase in metropolitan location. However, this may also be partly explained by the increased use of the American Indian response by persons of Latin American ancestry, a highly urbanized population. In 2000, 16 per cent of the census American Indian population (675,000 people) also reported a Hispanic origin. This was an increase from just 8 per cent of the 1990 Indian population (165,000 people). Data are not yet available for a formal assessment of the contribution of the components of change among natural increase, international immigration, internal migration, changes in metropolitan classification of space, and changes in reported racial identity.

However, one point that seems clear from these data is that the American Indian population, like other American populations, is now and will remain a primarily metropolitan population. At the same time, reservations and similar enclaves – largely, but not entirely located in relatively low-density non-metropolitan areas – continue to be home to an important component of the Indian population. This component continues to grow in absolute number, even as it declines as a portion of the Indian population.

What is the relationship between urban and reservation populations? The answer to this question must be segmented. A subset of the urban Indian population of unknown size does have continuing relationship to reservation populations. The nexus between reservation and city continues to be one component of American Indian migration flow. Systems of migration between reservation and city are relevant to a declining fraction of the urban Indian population, both because of the continuing growth of the population through increases in self-identification as an Indian by mixed ancestry Indians, and because of the increasing generational remove from the reservation experience of the children, grandchildren and great-grandchildren of previous generations of urban Indian migrants. However, the patterns of migration flow between reservation and city remain important to understanding the experiences of reservation populations.

There is important continuity in the structural dynamics affecting the choices of reservation Indians. The persistent tension between under-development of reservation economies and reservation-based tribal sovereignty that has existed through the entire history of the reservation has not been resolved for most tribes. This tension maintains for many reservation populations the same demographic system that has character-ized many reservations throughout the twentieth century: high rates of natural increase, low employment to population ratios, economic pressure to migrate to urban employment centres.

The past half-century has seen a significant narrowing of the differences in constitutive culture between reservation populations and other Americans. The gross cultural unfamiliarity reported for some urbanizing migrants in the 1950s is not likely to present the same kind of challenges today given rising rates of formal schooling, the pervasive influence of mass media, and the increase in social contact between Indians and non-Indians over this half-century. This is not to say that urbanizing migrants from reservation-to-city do not experience adjustment problems. Conditions of chronic unemployment in reservations may continue to make the transition into urban labour markets difficult for reservation Indians, as sociologists suggest that it does for residents of inner city communities (Wilson *et al.* 1995). These issues have recently received less attention from academic researchers than they once did.

It would be useful to revisit for current reservation populations many of the questions about rates of migration to, and adaptive responses within cities. The relationship between migration and economic opportunities, patterns of life course migration from and to reservations, the barriers to employment and social participation in mainstream institutions of reservation out-migrants are all topics that continue to deserve, even if they do not receive much, attention. Institutional, cultural, demographic and economic conditions affecting migration between reservation and cities have changed sharply in the past several decades, even while the economic underdevelopment of reservation communities continues to pose stark choices to reservation Indians about whether to stay or go.

Unfortunately, current data on these topics are scarce. Census data remain the only nationally representative data set available with which to study American Indian migration. Census data are limited by their cross-sectional rather than longitudinal character, and by the inability to identify subsets of Indians who have migrated from reservation areas, or who are likely to migrate to them. Studies that examined patterns of migration flow longitudinally for reservation communities would be extremely useful if we are to understand current migration patterns, even if these must be limited to studies of one or a few reservations.

Future directions

At the time of writing, economic data were not available from the 2000 census to assess how much conditions of economic underdevelopment and the migratory system that accompanies it had changed since 1990, and for how many tribes. If anything, data from 1990 showed retrogression from reservation conditions in 1980 (Gregory *et al.* 1996; Trosper 1996). Census data for 2000 will likely find that in the 1990s a few tribes successfully used casino gaming or other development initiatives to create fundamental changes in economic conditions on reservations. However, recent Bureau of Indian Affairs labour force estimates portray no fundamental change for most reservations.

Each passing decade continues to diminish the relative numerical importance of the reservation-to-city migration nexus to the experiences of most urban Indians. Even in the immediate post-World War II decades, urban Indian populations were probably much more diverse with respect to urban experience and personal characteristics than much of the contemporaneous literature implied (Garbarino 1971). Subsequent growth of the urban Indian population through changes in identification have decreased the fraction of the urban Indian population that has personal experience of reservation life and for whom reservation residence is a practical alternative. So too has cohort succession: some urban Indians are not the children, grandchildren or even great grandchildren of the immediate post-war migrants.

A substantial fraction of this urban American Indian population is now well-integrated into social and economic relations with other Americans. It is not clear to what extent the migration or other demographic behaviour of this subset of the Indian population follows a distinctively Indian pattern, rather than the patterns associated with the life cycle effect, family structure, and social class that pertain to the general population. This topic is again one that deserves further exploration but for which data are scarce. Census data remain the only data set that can be used to address these issues, but it is difficult to identify subsets of the Indian population using census data (Snipp 1989, 1996).

The most important study of the urban Indian experience in the 1990s was Weibel-Orlando's (1999) *Indian County, L.A.*, a study of the Indian community in Los Angeles, home to the nation's largest urban Indian population. This study marks a shift in emphasis from studies conducted two decades before, by emphasizing the social and institutional structures that sustain urban Indian community as a permanent if quantitatively minor part of Los Angeles' ethnic scene, rather than transitional experiences of new urban immigrants. Weibel-Orlando describes this community as a persistent structure inhabited by a changing mix of individuals who pass through the structures' position. Participation in these structures – for example, as leaders or active participants in Indian social service agencies, powwow clubs, churches – is volitional, often limited to a portion of an individual's life course. A relatively small minority of census identified Indians participate in such institutions. On this basis, urban Indian community institutions may be expected to persist on a more-or-less permanent basis, though subject to vagaries of funding. We have relatively little current information from Weibel-Orlando or others about the structural dynamics of individual recruitment into participation in these urban community structures, and the relationship between geographic migration and this participation.

One persisting question about urban Indian populations concerns their formal legal relation to tribal governments and to the United States government. Many urban Indians are not formally enrolled as members of federally recognized tribes, and have no specific claims to services or support from either tribes or the federal government. Many other urban Indians do meet tribal membership requirements – often based on attaining a given threshold of provable tribal ancestry – and may remain citizens of tribes for their entire life. The needs and interests of long-term off-reservation residents – even those who are enrolled tribal members – may diverge sharply from the needs of reservation populations. Urban Indian populations compete with reservation populations for a share of tribal resources and government spending that is targeted for the health and welfare of Indians.

The uncertain status of out-migrating reservation Indians remains a persisting legacy of the continuation of the reservation system. The economic pressures that led many tribes to shed a portion of each generation born on the reservation to off-reservation residence in adulthood remains in place. Old-fashioned assimilationist assumptions that such off-reservation Indians will eventually break off social relationships with their reservation cousins are unwarranted in an age of diminished costs of travel, diffusion of mass culture, and formalized administration of tribal enrolment rules. Nonetheless, it remains the case that the difference in context between reservation and urban settings may give rise to important differences between on- and off-reservation Indian populations. Among these differences is the relative inattention by both the tribal and federal government to urban populations outside of reservation areas.

Some social scientists have argued that certain segments of urban Indians have social welfare needs that are not well met by existing service agencies. Yet in competing for support to serve these needs, urban Indian organizations are often over-matched. They do not have the same status as "domestic dependent nations" that is accorded to tribal governments, and therefore do not have the same level of access to federal support as do tribal governments. Urban Indian organizations typically represent populations that are too small to compete effectively for resources against other ethnic populations for a share of government appropriations to serve the needs of disadvantaged populations. Thus the exact status of Indigenous Americans who are out-of-place with respect to the geography of the reservation system remains an unsettled point.

References

Ablon, J. (1964) 'Relocated American Indians in the San Francisco Bay Area: social interaction and Indian identity', *Human Organization*, 23: 296–304.
Barsh, R.L. (1994) 'Indian policy at the beginning of the 1990s: the trivialization of struggle', in L. Letgers and F. Lyden (eds) *American Indian Policy. Self Governance and Economic Development*, Westport, CT: Greenwood Press.
Bernstein, A. (1991) *American Indians and World War II: Toward a New Era in Indian Affairs*, Norman: University of Oklahoma Press.
Castile, G.P. (1998) *To Show Heart: Native American Self-Determination and Federal Indian Policy, 1960–1975*, Tucson: University of Arizona Press.
Cornell, S. (1988) *The Return of the Native: American Indian Political Resurgence*, New York and Oxford: Oxford University Press.
Eschbach, K. (1993) 'Changing identification among American Indians and Alaska Natives', *Demography*, 30: 635–52.
Eschbach, K. (1995) 'The enduring and vanishing American Indian: American Indian population growth and intermarriage in 1990', *Ethnic and Racial Studies*, 18: 89–108.

Fixico, D. (1986) *Termination and Relocation: Federal Indian Policy, 1945–1960*, Albuquerque: University of New Mexico Press.

Fixico, D. (2000) *The Urban Indian Experience in America*, Albuquerque: University of New Mexico Press.

Fuchs, L.H. (1990) *The American Kaleidoscope: Race, Ethnicity, and the Civic Culture*, Middletown, CT: Wesleyan University Press.

Garbarino, M. (1971) 'Life in the city: Chicago', in J. Waddell and O.M. Watson (eds) *The American Indian in Urban Society*, Boston: Little, Brown.

Gregory, R., Abello, A. and Johnson, J. (1996) 'The individual economic well-being of Native American men and women during the 1980s: a decade of looking backwards', in G. Sandefur, R. Rindfuss and B. Cohen (eds) *Changing Numbers, Changing Needs. American Indian Demography and Public Health*, Washington DC: National Academy Press.

Hackenberg, R. (1972) 'Reluctant emigrants: the role of migration in Papago Indian adaptation', *Human Organization*, 31: 171–86.

Hoxie, F. (1984) *A Final Promise. The Campaign to Assimilate Indians, 1880–1920*, Lincoln: University of Nebraska Press.

Kalt, J. and Cornell, S. (eds) (1992) *What Can Tribes Do? Strategies and Institutions in American Indian Economic Development*, Los Angeles: American Indian Studies Center, University of California at Los Angeles.

McDonnell, J. (1991) *The Dispossession of the American Indian, 1887–1934*, Bloomington: Indiana University Press.

Nagel, J. (1996) *American Indian Ethnic Renewal: Red Power and the Resurgence of Identity and Culture*, New York and Oxford: Oxford University Press.

Neils, E. (1971) *Reservation to City: Indian Migration and Federal Relocation*, Chicago: University of Chicago, Department of Geography.

Passel, J.S. (1996) 'The growing American Indian population, 1960–1990: beyond demography', in G. Sandefur, R. Rindfuss, and B. Cohen (eds) *Changing Numbers, Changing Needs: American Indian Demography and Public Health*, Washington DC: National Academy Press.

Passel, J.S. and Berman, P. (1986) 'Quality of 1980 Census data for American Indians', *Social Biology*, 33: 163–82.

Prucha, F.P. (1984) *The Great Father: The United States Government and American Indians*, Lincoln: University of Nebraska Press.

Shoemaker, N. (1999) *American Indian Population Recovery in the Twentieth Century*, Albuquerque: University of New Mexico Press.

Snipp, C.M. (1989) *American Indians: The First of This Land*, New York: Russell Sage Foundation.

Snipp, C.M. (1996) 'The size and distribution of the American Indian Population: fertility, mortality, residence and migration', in G. Sandefur, R. Rindfuss and B. Cohen (eds) *Changing Numbers, Changing Needs: American Indian Demography and Public Health*, Washington DC: National Academy Press.

Snipp, C.M. and Sandefur, G. (1988) 'Earnings of American Indians and Alsaka Natives: the effects of residence and migration', *Social Forces*, 66: 994–1008.

Sorkin, A.L. (1978) *The Urban American Indian*, Lexington, MA: D.C. Heath.

Thornton, R. (1987) *American Indian Holocaust and Survival: A Population History Since 1492*, Norman and London: University of Oklahoma Press.

Thornton, R., Sandefur, G. and Grasmick, H. (1982) *The Urbanization of American Indians: A Critical Bibliography*, Bloomington: Indiana University Press.

Trosper, R. (1996) 'American Indian poverty on reservations', in G. Sandefur, R. Rindfuss and B. Cohen (eds) *Changing Numbers, Changing Needs: American Indian Demography and Public Health*, Washington DC: National Academy Press.

Tyler, S.L. (1973) *A History of Indian Policy*, Washington DC: United States Department of the Interior.

US Bureau of the Census (1993) *1990 Census of Population: Social and Economic Characteristics. American Indian and Alaska Native Areas*, Washington DC: US Government Printing Office.

US Bureau of the Census (2001) *Census 2000 Redistricting Data. Public Law 94–171* (machine readable file). Washington DC: produced and distributed by the United States Bureau of the Census.

Waddell, J. and Watson, O.M. (eds) (1971) *The American Indian in Urban Society*, Boston: Little, Brown.

Waters, M. (1990) *Ethnic Options*, Berkeley: University of California Press.

Weibel-Orlando, J. (1999) *Indian Country, L.A. Maintaining Ethnic Community in Complex Society* (revised edition), Chicago: University of Illinois Press.

White, R. (1991) *"It's Your Misfortune and None of My Own." A History of the American West*, Norman and London: University of Oklahoma Press.

Wilson, F.D., Tienda, M. and Wu, L. (1995) 'Race and unemployment: the labor market experiences of Black and White men, 1968–1988', *Work and Occupations*, 22: 245–70.

5 Government policy and the spatial redistribution of Canada's Aboriginal peoples

James S. Frideres, Madeline A. Kalbach and Warren E. Kalbach

Canada's Aboriginal population comprises North American Indians, Métis and Inuit who collectively constitute a small (3.9 per cent), but growing share of the country's total population. This chapter opens with a brief overview of the historical basis of White settler contact with each of these groups as essential context for an analysis and explanation of contemporary patterns of Aboriginal interaction and mobility following Canada's demographic transition from a rural to an urban post-industrial society. Distributional patterns and other specifics of Aboriginal demography from the most recent Canadian census are examined as a basis for understanding changing Aboriginal mobility patterns and their consequences over the past century.

The data sources for this analysis are the national censuses of Canada that have been collecting and publishing data on ethnic and cultural origins of the population since before Confederation in 1867. Users of these data for research on the Aboriginal peoples of Canada have been well aware of the weaknesses and limitations, especially of the early censuses, which continue to affect the validity and reliability of ethnic ancestry and identity data. There are still difficulties in achieving complete coverage and comparable data over time, especially with respect to the Aboriginal population, because of changes in the question wording, format, examples, instructions and data processing, as well as the social environment prevailing at the time of each census (Statistics Canada 1996).

Major factors contributing to the persistence of these problems have included a shift from traditional census interviewing procedures to self-enumeration questionnaires since the 1971 census. Also of importance has been the change from patriarchal determination of single ethnic ancestry, to recognition of the legitimacy of multiple ancestries, and emerging new "identities" among second and later generations, though these may not always correspond to the ethnic and cultural origins of earlier ancestors because of intervening socialization and assimilation experiences.

The historical analysis in this paper is primarily based on the response data obtained from the traditional ethnic origin census question: "What is the ethnic origin of your ancestors?" Census methods for collecting these data, periodically criticized for their possible effect on the validity and comparability of the data, have continued to be problematic for the population in general, and particularly troublesome with respect to the problems of enumerating and quantifying Canada's Aboriginal peoples. In 1996, efforts were made to satisfy the government's growing need for better and more comprehensive data on Aboriginal people. In particular, a new question was introduced in order to allow Aboriginal people to distinguish between ethnic and cultural origins as ancestry, as provided for by the ethnic origin question, and what they consider to be their current Aboriginal identity by personally identifying with at least one of the recognized Aboriginal groups – North American Indian, Métis, or Inuit.

The major effect of these changes in the 1996 census enumeration procedures has been to shift the focus away from the ancestral background of Aboriginals to their own perception of their "identity" in contemporary Canadian society as affected by their collective and individual integrative and assimilative experiences. These "self response" Aboriginal "Identities" have been incorporated into this analysis of Aboriginal characteristics and current mobility patterns in Canada.

Historical movement of Aboriginal peoples

This section describes the spatial movement of Indians, Inuit and Métis from the eighteenth century to the end of the twentieth century. The mobility of these three groups reveals similar patterns of redistribution in that few have moved far from their traditional territories and places of birth. At the same time, all groups have undergone a process of urbanization as reflective of the country as a whole.

North American Indians (First Nations)

Initial contacts between the First Nations of Canada and Europeans during the sixteenth and seventeenth centuries impacted on both the social and physical components of Indian culture and in doing so, influenced their place of residence (Patterson 1972). Traditionally, Canada's North American Indians were hunters and gatherers and hence, moved "in whole or partially, at various seasons of the year, in search of economic opportunities or social interactions" (Miller 1999: 6). Long-term settlement tended to be associated with a good source of game and trade. As the European presence gained momentum throughout the eighteenth century, one of the first consequences was greater social interaction among Indian peoples, which then spawned cultural interchanges among various bands that had never been experienced before.

For example, Cree Indians moved from the Hudson Bay area westwards to the foothills of the Rocky Mountains between the late eighteenth and early nineteenth centuries. The Ojibwa moved gradually southeast into southern Ontario during the first half of the eighteenth century, creating conflict between them and Iroquois residents (Patterson 1982). Other examples of early movement included the Abenaki who moved into Quebec from New England in the seventeenth century, and the Iroquois and Delaware who migrated into southern Ontario in the late eighteenth century. Elsewhere, the Chipewyan and Cree moved north and south of their traditional Ontario hunting areas, while the Blackfoot of the prairie also moved further west into southern Alberta. One consequence of these movements and intrusions into others' territories, was the establishment of confederacies by means of loosely coupled associations between nations. For example, the Ottawas, Ojibways and Potawontdnis forged a Council of Three Fires, while the Mohawk, Cayuga, Oneida, Onondaga, Seneca and Tuscarora formed the Six Nations.

As inter-Indian contact continued, the flexibility and adaptability of Indian culture made it possible for Indian peoples to retain important aspects of their tradition and culture. In short, they adapted to changing conditions while maintaining their identity and sense of community. However, it would not be until late in the twentieth century that a sense of "Pan-Indianism" would emerge.

As the eighteenth century drew to a close, European settlers began to colonize Canada and the settlement patterns of North American Indians changed forever. To clear the land for these new arrivals, the government persuaded them to "take treaty". This action of treaty making established the right of the state to claim ownership of vast tracts of land while at the same time making provisions to establish reserves (or homelands) – small areas of land set aside for the semi-exclusive use and residence of North American Indians. As treaties and land transfers proceeded (except in the northern regions of the country), increasing numbers of Indians were forced to remain within the confines of these homelands. Figure 5.1 illustrates the spatial extent of this land surrender, a process which extended well into the twentieth century (Kydd and Kredentser 1992). The figure reveals that treaties established between the Indians and the federal Crown took place over many years. It also shows that large parts of Canada were never "treatied" out. That is, Indians living in major portions of Canada have never established a treaty with the Crown. What is not shown on the map are large areas of land that have been "surrendered" by Indians through modern day treaties, or "comprehensive claims", such as found in western Northwest Territories, eastern and western Arctic and Northern Quebec. Finally, the map reveals the current major cultural groupings of Indians in Canada and identifies their spatial location across Canada.

To ensure an orderly transfer of land and to monitor/control the behaviour of Indians, the Indian Act, created in 1876, gave jurisdiction to the federal government to intervene in all aspects of their lives. The reserves were lands set aside by the federal government that settlers (or the

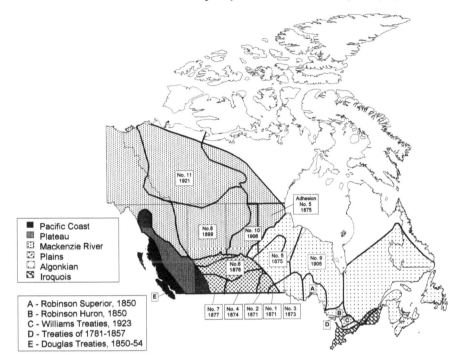

Figure 5.1 Indian land surrendered for treaties, up to 1921.

Source: Ethnic Studies and Population Research Laboratory, The University of Calgary.

government) had little interest in occupying or using. Hence, today many of the reserves still remain isolated, occupy poor agricultural land, and contain few natural resources that can be exploited.

By the twentieth century it was clear Indians could not contain the forces of a settler society. At the same time settlers were convinced that it was their right to colonize Canada and Indigenous populations would have to assimilate or disappear. Subsequent actions taken by settlers to deal with Canada's North American Indians left no doubt that they were the dominant force and could, to a great extent, dictate the life-forces impacting upon Aboriginal peoples. Numbered treaties (the early treaties were named after the principal negotiator while later treaties were simply numbered one to eleven) were designed to take away the land of Aboriginals. In addition, laws detrimental to Indian culture were enacted, such as the anti-potlatch laws of the northwest coast, the banning of prairie Indian dances, and outlawing of various Indian ceremonies, to ensure that the dominant group in Canada could effectively control the actions of Aboriginal peoples. Moreover, Aboriginal people were effectively excluded from participation in the new agricultural economy. For example, lobbying efforts in the late nineteenth and early twentieth centuries by farmers kept Indian Affairs from providing any support for Indians to develop their agricultural base. In other examples of social control, Indians were excluded from voting in provincial or federal elections until 1960. New attitudes and beliefs about Aboriginal people provided a further rationale and justification for actions taken against them. For example, notions of racial superiority were placed within a religious, scientific, legal and systemic context that would weave the legitimacy of those attitudes and beliefs into the social fabric of Canadian society. Arguably, this still prevails.

By the early twentieth century, the North American Indian population had been reduced in size from an estimated 250,000 at the time of first contact to around 120,000. It was geographically isolated in reserve areas, was segregated from the emergent economic shift towards industrialization and modernization, and was believed by governments of the time to be destined for extinction, both physically and culturally. By this time, the Beothuks of Newfoundland had become extinct after extensive genocidal action by settlers, and so there was precedent and reality in such beliefs. Thus, the first two centuries of contact between North American Indians and settlers may be characterized as falling into two broad phases. The first was initially episodic in nature and the relationship between early colonists and resident Indigenous peoples could be characterized as symbiotic. However, as Europeans continued to colonize the country, Indians were drawn deeper into the Euro-Canadian sphere of influence. As a result, they were forced into an economic dependency and experienced major irreversible cultural changes. The second phase may be described as the reserve period (1850–1930), when Indians were forced by treaties to live in geographical enclaves and were subjected to extreme political and social control.

In the latter half of the twentieth century North American Indians began to regain some political and social control of their lives. Between 1950 and 1970, the assimilation policy that had been so forcefully pursued by government until this time, was wound back and a more integrative phase emerged. The assimilation period focused on making Indians take on all the social and cultural attributes of the dominant society. Through this policy it was believed that Indians would give up their culture and replace it with another. Integration focused on developing strategies by which Indians could retain their culture and actively participate in the social, political and economic institutions of the larger society. The contemporary phase of Indian policy commenced in the 1970s when the Supreme Court of Canada first recognized "Aboriginal Rights". Since then, North American Indians have been granted increasing means to self-government and self-determination. Over the past three decades, through a variety of policies and court cases, the Government of Canada has agreed to allow Indians to begin to take some limited control of their lives and communities. This control is largely over programme expenditures and reflects the devolution of centralized federal control and the withdrawal of policy that promoted an urbanization process.

Since the 1973 Supreme Court decision (*Calder* v. *Attorney General of British Columbia*) confirmed the existence of Aboriginal rights, the Federal government has embarked upon a policy of devolution, attempting to move reserve control over to local residents. Today, nearly one hundred negotiations regarding devolution are underway with Indians, impacting nearly 500 First Nation communities. Both land and treaty agreements have been made with a number of groups such as the Sechelt (B.C.), Cree-Naskapi (Quebec), Mi'Kinaq (Nova Scotia), and several First Nation groups in Yukon. As a result, by the end of the 1990s bands controlled about 85 per cent of all expenditures on reserves. At the same time, over the same decade, nearly 500,000 hectares of land was converted into reserve land because of specific claims won by Indians in their attempts to improve their lives by reducing poverty and raising the standard of living of those living on reserves. To the extent that these policies are effective, Aboriginal people on reserves will have fewer incentives or reasons to move locally or migrate to the large urban centres.

At various historical junctures, then, Canadians have taken action towards North American Indian people based on the ideology of the time. The early belief was to segregate them and hope they would die out as an ethnic group. Later, attempts were made to assimilate them into Canadian society. More recently, there has been a strategy of integration into the social and economic structure of the wider Canadian society. Today there is some understanding by Canadians of the nature of oppression experience by North American Indians and a modicum of support for self-government is now underway. It is important to remember is that each of these strategies has emerged from unilateral decisions of the dominant society in imposing its beliefs and policies. Alongside this, the increased urbanization of First Nations peoples has been one consequence of their growing integration into mainstream Canadian social and economic life. While First Nations reserves were initially remote rural enclaves, and many remain so, enhanced transportation routes as well as the growth of cities have transformed some of them into communities with close connections to major urban areas. Some reserves, such as the Mohawk Reserve in Montreal, have been encapsulated by large urban areas, while the Tsuu Tina Reserve near Calgary, now finds itself adjacent to the nearby city of Calgary as the city limits have spread outwards. In short, many "rural" Indians have become urban without ever moving. Today, there are around 600 registered bands resident on some 2,400 reserves across the country, which collectively provide a home for roughly one-half of the Indian population.

Inuit

The Inuit traditionally lived in small camps in Canada's far north and the location of these camps changed depending on the seasons (Dorais 1999). As subsistence hunters and gatherers the pattern and timing of this movement was essentially determined by those places where game and fish were most plentiful at the time. Generally, these migrations were to the same places year after year and their movements were restricted to within dialect groups forming social and spatial boundaries which were rarely crossed (Dorais 1999). However, Inuit mobility decreased with the establishment of trading posts and missionary outposts and the Inuit eventually became semi-nomadic. The creation of trading posts meant that regular commercialized transactions developed between the Inuit and businessmen, and if a variety of basic services were also attached to trading posts and made available to the Inuit, then their mobility became further constrained by these new attractions. The introduction of a money economy to the region, followed by the subsequent provision of social services and welfare, consolidated this process of attraction to settlements and resulted in almost complete sedentarization of the Inuit by the late 1950s.

Prior to the 1950s the Canadian government had, for the most part, left the Inuit alone, but the next three decades were different. The Department of Indian and Northern Affairs relocated an estimated 1,000 of Canada's Inuit between the 1950s and 1970s. Such forced relocations could be quite extreme. For example, the most widely known case occurred in 1953 when seven Inuit families from Quebec were moved from their homes in Port Harrison to Grise Fiord and Resolute Bay in the High Arctic over 1,900 kilometres away. They were joined by an additional 35 people, two years later. The rationale behind this forced migration appeared political – Canada was concerned about a potential weakening of its claim to Arctic sovereignty during the Cold War between the United States and Russia, and it was felt that Canada needed permanent residents in the High Arctic to strengthen its claim over the territory (National Film Board 1995).

The Inuit families who were moved to Resolute Bay and Grise Fiord found the climate in the High Arctic to be much harsher than in Port Harrison. They had no means by which to earn an income, food was scarce and fines were imposed for hunting more than one cariboo and for hunting muskox. Fishing was virtually non-existent due to the nearly always frozen Arctic Sea. The snow of Grise Fiord was unsuitable for igloo building so the Inuit had to live in canvas tents for the first year. In summary, the forced removal of the Inuit to the High Arctic imposed undue hardships on the people. While those involved were promised that they could return home if the move proved unsatisfactory, it was 30 years (1983) before they were returned to their ancestral homelands.

After World War II, as federal schools and health care facilities were established, the Inuit were encouraged to settle in a few permanent villages. By the end of the 1960s only a few traditional camps still existed in Canada's Arctic and by the 1970s most of the Inuit lived in villages, except in the Bathurst Inlet area. Inuit now live in permanent wooden houses instead of igloos and sod huts. By 1991 there were 46 Inuit communities in the High Arctic and most of these have outlying camps which the Inuit frequently move to for short periods of time to engage in traditional forms of hunting and fishing. Taylor (1996) found in his analysis of the Aboriginal Peoples Survey that just over half of all Inuit respondents had spent some time away from their usual place of residence in the previous year. While 60 per cent were away for a short time (four weeks or less), nearly one-third were away between five and twenty weeks.

The establishment of Nunavut, an Inuit Territory, in 1999, marked the emergence of Inuit self-government and Inuit-run organizations (Dorais 1999). Future local and migratory mobility rates for the Inuit will depend to a great extent on the impact of federal government policies aimed at sedentarization and modernization, as well as those emanating from the practice of Nunavut self-government.

Métis

Métis is a French word for "mixed" referring to the offspring of inter-marriage between European men and Indian women. The Métis have been particularly associated with Canada's Prairie Provinces, and especially the Red River Settlement in Manitoba, but intermarriage between Aboriginal and other Canadians has always occurred throughout the country (Gabriel Dumont Institute 1994). North American Indians often camped adjacent to trading posts so there was a great deal of contact between North American Indians and Europeans and considerable intermingling of the two groups leading to both legal and informal family arrangements that produced offspring. In fact the French encouraged intermarriage with North American Indians as a way of facilitating peace and harmony. These small settlements came about due to the fact that Europeans living at trading posts depended upon the local Indian populations for food, while the Indians profited from their association with the businessmen. The offspring of Métis often became mediators between the businessmen and Indians at trading posts, being both bicultural and bilingual (Dickason 1999).

The Métis of the early days, like their other Aboriginal counterparts, were hunters, gatherers and fishermen. They were semi-nomadic, moving only when the need for food and trading was necessary. A few of the Métis were also subsistence farmers, especially those established in the Red River area in the province of Manitoba. This group could be more accurately described as being "semi-sedentary" (Dickason 1999). During the early twentieth century the Métis continued to live in difficult circumstances, but in 1938, to help alleviate their problems, the Canadian government passed the Métis Population Betterment Act, making provisions for ten Métis settle-ments in northern Alberta (of which there are still eight today) (Dickason 1999) and two in British Columbia. Hence, for the past several decades the Métis have been relatively well established in these northern settlements. However, only a small proportion of Métis now live on Métis settlements.

With the exception of the ten settlements referred to, Métis have no land base and are consequently the most urbanized of the three Aboriginal populations.

Twentieth-century demographic revival

At the time of first contact with non-Aboriginal settlers in the sixteenth century, it is estimated that the Indian population was approximately 250,000. By the late nineteenth century, this figure had been reduced to just over 100,000 as a consequence of the introduction of European diseases and hostilities between the French, British and various Aboriginal groups. After this initial period of contact, sporadic conflict, and settlement, high levels of mortality among Indian peoples began to decline more rapidly than their high levels of fertility. As a consequence, a revival of population growth commenced in the 1920s, and has continued to the present time.

By the end of World War II, the Indian population had reached 160,900 and was increasing at about the same rate as the non-Indian population. By the 1960s the Indian growth rate had surpassed that of the rest of the population, enabling the population to recover to what it had been at the time of first settler contacts. Even more striking has been growth in the population claiming Aboriginal ancestry. This was estimated to have reached 220,000 in 1961, and almost half a million by 1981. By 1996, over one million Canadians were reported by the census to be wholly or partially of Aboriginal ancestry (Table 5.1).

Caution is advised in interpreting these census data as they vary considerably in their degree of completeness and comparability. For example, while much of the increase in the combined Aboriginal (Indian, Inuit and Metis) population can be attributed to a growing excess of births over deaths, the government's introduction of Bill C-31, in 1985, which

Table 5.1 Numerical and percentage distribution of Aboriginal and total populations: Canada, 1981–1996

	Canada total (millions)	Aboriginal		Aboriginal group		
		Total	%	North American Indian	Métis	Inuit
1981[a]	24.1	491,500	2.0	367,800	98,300	25,390
1986[b]	25.1	373,265	2.8	286,230	59,745	27,290
1991[a]	27.0	470,615	1.7	365,375	75,150	30,085
1996[b]	28.5	1,101,960	3.9	867,225	220,740	49,845

Source: Statistics Canada, 1981–1996 censuses of Canada.

Notes:
[a] Includes single and multiple origins.
[b] Includes single origins only.

redefined the criteria for claiming Indian status under the Indian Act, provided a new source of growth through "ethnic mobility". Individuals and their dependants previously disenfranchized were now allowed to reclaim their Indian identity and become legal Indians under the new federal Indian Act. As of 1997, well over 100,000 people had been given Indian status under the new Act, and the total count of Canada's population with single or partial Aboriginal ancestry rose from about three-quarters of a million in 1986 to a little over one million by 1996 (Government of Canada 1999). As a consequence of rising rates of natural increase and additions through ethnic mobility, the Aboriginal population doubled its share of the total population from approximately 2 per cent in 1981 to 4 per cent in 1996. This ethnic mobility has significant consequences for the interpretation of spatial redistribution as much of the apparent change in distribution of the Indian population (especially in favour of urban areas) could simply reflect the reclassification as Indian of individuals already based in urban centres, as pointed out by Clatworthy (1996).

In response to government requirements for new data regarding Canada's increasingly visible minority groups, the 1996 census followed the standard "ethnic or cultural" ancestry group question with an "identity" question, asking whether the person identified as a Native Indian, Métis or Inuit. Of the approximately 800,000 who identified themselves as having an Aboriginal identity, about two-thirds claimed they were Native Indian, a little over one-quarter indicated Métis, 5 per cent claimed Inuit identity, while the remainder indicated one or more of the three Aboriginal identities.

Regional distribution and movement of Aboriginal people

The net results of historical forces underlying the shifting Aboriginal and non-Aboriginal population movements are revealed in the 1996 census of Canada. The largest numbers of Aboriginal people are found in Ontario and the western provinces, which together account for 81 per cent of their total. Thus, Ontario accounts for the largest share with 18 per cent, followed by British Columbia (17 per cent), Manitoba (16 per cent), Alberta (15 per cent) and Saskatchewan (14 per cent). Of the remaining provinces and territories, Quebec accounted for 9 per cent, the Northwest and Yukon Territories 6 per cent, and the Atlantic provinces 5 per cent.

Comparing the proportions of Aboriginal ancestry in each provincial population, to the national average of 3 per cent, reveals an under-representation of Aboriginal ancestry in Ontario and all the eastern provinces. In contrast, there is over-representation and a greater concentration of Aboriginal ancestry in the western provinces and territories. The greatest relative concentrations of populations with Aboriginal ancestry are found to be in the Northwest and Yukon Territories where proportions of 62 and 20 per cent are recorded respectively, followed by Manitoba and Saskatchewan with 12 and 11 per cent.

Urbanization

Historically, the relatively greater concentration of Aboriginal people in the more rural regions of Canada suggests that they have been less affected by urbanization than the rest of the population. As recently as 1961, only 13 per cent of the Aboriginal population resided in urban areas compared to 70 per cent for the total population. However, during the decade 1961–1971, the increase in the rate of urbanization for the Aboriginal population was greater than that recorded for the total population. Already highly urbanized, the share of the total population resident in urban areas increased by just 10 per cent, compared to a substantial 138 per cent increase for the Aboriginal population whose level of urbanization increased from 13 to 31 per cent, as shown in Table 5.2. While there continued to be little change in the relatively high level of urbanization during the 1970s for the total population, urbanization of the Aboriginal population continued to increase, reaching 42 per cent by 1981 while the level for the total population remained relatively stable at around 76 per cent.

With the higher levels of urbanization achieved during the 1980s, a different measure of rural–urban distribution was needed for the 1991 census to better reflect the changing nature of urbanization occurring in Canada. Rather than the old rural–urban distinctions, population distributions have subsequently been compared for census metropolitan and non-metropolitan areas (CMAs and non-CMAs). While not comparable to earlier measures, metropolitan/non-metropolitan comparisons nonetheless still give an indication of the relative degree of urbanization achieved by

Table 5.2 Total and Aboriginal populations by rural and urban residence, 1961–1981, and by non-census and census metropolitan areas (CMA) of residence, 1991 and 1996

| | *Total population* | | *Aboriginal "ancestry" population* | | | |
| | | | *(Multiple origins response)* | | *(Single origin response only)* | |
	% Rural	*% Urban*	*Rural*	*Urban*	*Rural*	*Urban*
1961	30	70	n.a.	n.a.	87	13
1971	24	76	n.a.	n.a.	69	31
1981	24	75	28	72	58	42
	Non-CMA	*CMA*	*Non-CMA*	*CMA*	*Non-CMA*	*CMA*
1991[a]	23	77	39	61	74	26
1996[b]	39	61	51	49	71	29

Source: Statistics Canada census, 1961, 1971, 1981, 1991 and 1996.

Notes:
[a] 1991 and 1996 census both reported Aboriginal origin/ancestry data.
[b] "Self-reported" Aboriginal identity data for the population 5 years and over.

Canada's various Aboriginal populations, especially with reference to their attraction to the larger urban centres.

Groups with multiple ancestries are the product of past intermarriage and would be expected to show a greater degree of integration and cultural assimilation into mainstream society than those of single Aboriginal ancestry or identity. As shown in Table 5.2, 72 per cent of the multiple ancestry/identity population resided in urban areas in 1981 compared to only 42 per cent of the single ancestry/identity population. The equivalent figure for the total population was 75 per cent. In 1991, multiple ancestry/identity Aboriginals were still relatively more urbanized than those of single ancestry (61 per cent urban compared to 39 per cent). Although the index of urbanization was changed (from per cent urban to per cent CMA population), those with multiple Aboriginal identities were still more urbanized in 1996 (49 per cent), and more similar to the total population in this respect than the single origin/identity Aboriginal population.

These data for 1991 and 1996 show that the Aboriginal population as a whole, whether based on "ancestry" or "identity" data, was considerably less urbanized than the total population. Furthermore, the separate Aboriginal groups varied significantly from each other in this respect. As may be seen in Table 5.3, the Inuit were the least urbanized, with only 6 per cent in Canada's census metropolitan areas, followed by Native Indians with 26 per cent. In contrast, 41 per cent of the Métis and almost half (49 per cent) of Aboriginal people with multiple origins, reported metropolitan area residence. When Bill C-31 was introduced, over 100,00 non-status Indians (and their dependants) were converted into status Indian, some permanently, others until they turned 18 years of age. Nevertheless, it removed a large proportion of people previously reporting Métis or claiming multiple Aboriginal origins. Only the Inuit would not have been impacted by Bill C-31. This legal redefinition could easily have produced a pronounced impact on the proportion living in urban areas. Hence, the actual amount of residential shift due to migration could have been minimal. However, at the time of writing, it was not possible to ascertain the extent to which this residential shift was a function of physical mobility or "ethnic mobility". Notwithstanding the lack of comparability between the "ancestry" and "identity" definitions, the data in Table 5.3 show modest but consistent evidence of increased urbanization for each of the three Aboriginal groups between 1991 and 1996.

A stronger net urban movement of Native Indians appeared after World War II when a federal policy of "democratization" was introduced. This allowed Indians to more fully participate in mainstream social and political institutions by allowing them to vote in provincial and federal elections, and to become more active in the economic sphere. For example, Indian agents no longer needed to be consulted regarding economic activities on reserves. At this time, each reserve had an Indian Agent (a representative of the federal government) who was responsible for the operations of the

Table 5.3 Percentages of Aboriginal "ancestry" population, 1991, and "identity" population, 1996: non-metropolitan area and census metropolitan area place of residence

	Non-metropolitan area		Census metropolitan area	
	1991	1996	1991	1996
Total population	23	40	77	60
Total Aboriginal[a]	74	71	26	29
Native Indian single identity	75	74	25	26
Métis single identity	60	59	40	41
Inuit single identity	95	94	5	6
Multiple Aboriginal identities	n.a.	51	n.a.	49

Source: Statistics Canada, 1996 census of Canada, Public Use Microdata File, Individual File.
Note:
[a] 1991 census data based on the 'Ethnic origin/ancestry' question, 1996 census data based on the Aboriginal identity question. Note that Aboriginal identity data were not collected in 1991.

reserve by ensuring that the Indian Act was properly applied, and that resident Indians conducted their activities in accordance with the Act. For example, permission was required from an Agent before agricultural goods could be sold off-reserve, while Indians wanting to leave the reserve had to secure a pass from the Agent.

More integrative policies were implemented in other spheres, such as education, to ensure higher levels of social and economic participation. The means to this was to focus service delivery on towns and not on reserves. For example, federally controlled reserve schools were phased out of existence and students were forced to attend provincial schools off-reserve if they wanted to obtain a formal education. During the period 1940–1960, the federal government closed down many schools on reserves and forced Indian children to attend the nearest provincial schools. Moreover, because there were few secondary schools on the reserves, attendance at a high school required students to relocate to a provincial school. In many cases the nearest provincial school was a considerable distance from the reserve so if parents wanted to be near their children while they attended school (made compulsory by the state), then whole families often had to relocate. Other policy initiatives that served to stimulate movement off reserves included the construction of all-weather roads, the slowing down of housing construction on reserves, and active promotion of the urban way of life. A natural consequence of such policies was a period of more rapid urbanization for Native people. Today, almost one in three Aboriginal people live in a large metropolitan areas, while some 25 per cent live in other, smaller urban areas.

In a seminal study of the community characteristics associated with migration patterns (Gerber 1984), a number of community attributes related to out-migration rates from reserves were explored. Seven structural factors in particular were found to provide substantial explanation for the rates observed. These included close proximity to urban centres, low levels of institutional completeness, low rates of fertility, access to good all-season roads, high levels of personal capital, and individualistic, as opposed to communal, orientation. Females were also found to be more inclined than males to migrate. Such factors continue to influence the rate of urbanization, and government policies are directed toward influencing these in order to promote urbanization. However, the urbanization of Indians today does not reflect the traditional step-wise pattern of a rural population moving to small towns, and then on to urban areas; rather it involves a direct movement from reserves to large metropolitan centres.

In reality, the process of urbanization for Indians is complex and is likely to occur in a series of stages depending upon the ability of individuals to successfully participate in the institutional network of an urban centre. For example, successful graduation through a public service organization increases the likelihood of becoming a permanent urban resident. However, relatively few Indians are able to achieve in this way and thus tend to lack the requisite social and technical skills for sustained public sector employment. Even member organizations, such as Indian Friendship Centres and Urban Indian Associations, cannot adequately fulfil this role as they often lack the staff or organizational capacity to provide requisite training and experience. In effect, current organizational structures in urban centres are ill-equipped to deal with Indian settlement issues. As such, relatively few Indians are successful adapters to the urban milieu and they tend to assume sojourner status as they move in and out of urban centres (Reeves and Frideres 1981).

The net result of this urbanization of the Aboriginal population for seven of the larger Canadian metropolitan areas is indicated in Table 5.4. Overall, these seven CMAs account for some 15 per cent of the Aboriginal population in 1996. However, the impact of urbanization as measured by the Aboriginal proportion of each CMA population varies somewhat from the central Canadian cities of Winnipeg, Regina, Saskatoon and Edmonton which have a relatively high proportion of Aboriginal population, to the larger coastal and eastern cities such as Vancouver and Toronto with relatively small proportions.

As the status Indian population has shifted towards urban residence, a commensurate decrease in the "on reserve" population has occurred. Today, it is estimated that just over 58 per cent of the status Indian population is resident on reserves. Regional variations run from a high of 74 per cent for the Northwest Territories, to a low of 51 per cent on Ontario's reserves. The overall data from 1930 to the present, showing a net migration flow off the reserves, is further evidence of a continuing

Table 5.4 Aboriginal populations[a] in selected census metropolitan areas: Canada, 1996

Census metropolitan area	Total Aboriginal population	Aboriginal population as % of total population
Toronto	16,100	0.4
Winnipeg	45,750	6.9
Regina	13,605	7.1
Saskatoon	16,160	7.7
Calgary[b]	15,200	1.9
Edmonton	32,825	3.8
Vancouver[b]	31,140	1.7

Source: Statistics Canada, *The Daily*, 13 January 1998, p.5

Notes:

[a] The population identifying themselves as either North American Indian, Métis or Inuit.

[b] These CMAs contain, within their boundaries, Indian reserves which were incompletely enumerated during the 1996 census. Totals for these CMAs as well as for their Aboriginal populations would be affected by this under-enumeration on their Indian reserves.

and perhaps quickening pace of urbanization among Native people in Canada.

Increased population mobility and ethnic intermarriage are known concomitants of urbanization in North American industrializing societies. Therefore, it is not particularly surprising or unexpected to find that mixed ancestry or identity groups, such as the Métis and those Aboriginals reporting multiple ancestries or identities, are more likely to be found in metropolitan than non-metropolitan areas. Metropolitan area data would also be expected to show the relatively greater residential mobility which has come to characterize those living in metropolitan areas in contrast to those continuing to live in non-metropolitan and more rural areas.

The increasing urbanization of the population, in general, implies a higher degree of residential mobility simply as a result of more people moving to the larger metropolitan centres. By 1961, Canada's population was exhibiting a relatively high level of mobility with almost half (45 per cent) indicating a change in residence over the previous five years (Table 5.5). While the total population mobility rate declined from 45 to 41 per cent between 1961 and 1996, Canada's Aboriginal population experienced a substantial increase in their mobility from 45 to 55 per cent. During this period of relatively more rapid urbanization of the Aboriginal population, it is not surprising that their overall mobility rates should have increased to this extent. The actual increase might have been even higher had it not been for the fact that many live on reserves either within or near to large urban centres, as in the case of Montreal, Calgary and Vancouver. In such cases, where it is not necessary to move in order to seek employment off the reserves, urbanization would tend to have reduced residential mobility.

Table 5.5 Mobility status for Aboriginal and total population: Canada, 1961 and
1996

		Non-movers	Movers (%)		
			Total	Local	Migrants
Total population	1961	55	45	25	20
	1996	59	41	24	17
Aboriginal	1961	55	45	33	12
	1996	45	55	33	22

Source: Dominion Bureau of Statistics, 1961 Census, Special Tabulations; Statistics Canada,
1996 census of Canada, Public Use Microdata File, Individual File.

Note:
Includes Native Indian, Inuit, Métis and multiple Aboriginal origins in 1996, but only the
Native Indians and Eskimo (Inuit) in 1961.

Local movers and migrants

Differences in total mobility rates between Aboriginal and non-Aboriginal
populations have obscured important variation in other types of mobility,
such as local movement (within an urban or surrounding area) and longer
distance movement (from one urban centre to another, or from a rural to
urban residence). It is clear from Table 5.5 that the majority of the population did not change residence during either the 1956–1961, or the
1991–1996 period. It is also clear that for those who were movers at some
time during this 35 year period, their local mobility was greater than their
longer distance mobility. Total mobility rates for the Aboriginal groups
increased from 45 per cent in 1961 to 55 per cent in 1996, while they
declined for the total population. For both populations, local mobility rates
exceeded longer-distance migrant rates, and although they were significantly higher for the Aboriginal population in 1996 (33 per cent compared
to 24 per cent), they showed little change during the period since 1961.
However, while Aboriginal local mobility rates were considerably higher
than those for the total population, and had remained constant at 33 per
cent, the migrant mobility rate for the Aboriginal population almost
doubled, increasing from 12 per cent in 1961 to 22 per cent in 1996.
Clearly, this increase in total mobility experienced by the Aboriginal
population, was due entirely to an increase in their longer-distance migration. Again, this contrasted significantly with the total population, whose
total mobility had declined from 45 to 41 per cent, primarily as a result of a
decline in the rate of longer-distance migration from 20 to 17 per cent.

As indicated in Table 5.6, all population groups in Canada's metropolitan areas displayed higher total mobility rates than those living in the
non-metropolitan areas. The difference between metropolitan and non-
metropolitan mobility rates was most marked among those reporting
specific Aboriginal identities. Overall, 67 per cent of Aboriginal people in

Table 5.6 Mobility status for total, Aboriginal and non-Aboriginal populations:
Canada's census metropolitan and non-metropolitan areas, 1996

Aboriginal group	Total population 5 years+ (n)	Non-movers %	Movers (%)		
			Total	Local	Migrants
Total population					
Total Population	25,654,824	59	41	24	17
Non-Aboriginal	24,698,340	59	41	24	17
Total Aboriginal[a]	686,484	45	55	33	22
Native Indian	465,948	47	53	32	21
Métis	182,412	42	58	34	24
Inuit	32,400	40	60	45	15
Multiples	5,724	38	62	33	29
Census metropolitan areas					
Total Population	15,223,752	56	44	27	17
Non-Aboriginal	15,020,712	57	43	27	16
Total Aboriginal[a]	203,040	33	67	43	24
Native Indian	123,156	32	68	43	25
Métis	75,168	34	66	45	21
Inuit	1,908	34	66	23	43
Multiples	2,808	35	65	41	24
Non-census metropolitan areas					
Total Population	10,431,072	62	38	20	18
Non-Aboriginal	9,947,628	63	37	19	18
Total Aboriginal[a]	483,444	51	49	29	20
Native Indian	342,792	52	48	28	20
Métis	107,244	48	52	27	25
Inuit	30,492	41	59	47	12
Multiples	2,916	41	59	26	33

Source: Statistics Canada, 1996 census of Canada, Public Use Microdata File, Individual File.
Note:
[a] Refers to those "self-reporting" one or more of the three Aboriginal identities.

metropolitan areas reported a change of residence between 1991 and 1996
compared to only 49 per cent of their counterparts in non-metropolitan
areas. The equivalent figures for the non-Aboriginal population were 43
and 37 per cent respectively.

The corresponding mobility rates for all non-Aboriginals were 24 per
cent for local movers and 17 per cent for migrants. In general, mobility
rates for both Aboriginal local movers and migrants were higher in the
metropolitan areas than the non-metropolitan areas, although some
variation to this trend is observed among the Inuit. In each case, the

mobility rates for Aboriginal movers were also higher than for non-Aboriginal movers, except for Inuit in non-CMAs. Further differences in mobility are observable within the population indicating Aboriginal identities, between Native Indians, Métis, Inuit, and Aboriginals with multiple identities. Those with multiple origins, and the Inuit reported the highest mobility rates of 62 and 60 per cent respectively, closely followed by the Métis with 58 per cent and 53 per cent for the Native Indians Among these groups, the Inuit exhibited the highest local mobility, 45 per cent, compared to just 33 per cent for the other three Aboriginal identity groups.

Conclusion: government policy and mobility

The Aboriginal population of Canada survived the impact of settlement by Europeans, but barely so. By the early twentieth century, the Aboriginal population had been substantially reduced in size, more or less confined to isolated land areas, called reserves, and segregated from mainstream social and economic trends towards urbanization, industrialization and modernization. Quite simply, they were on their way towards extinction, both demographically and culturally. The establishment of reserves during the years of treaty-making, forced Aboriginal people to live in geographical enclaves where they were subjected to extreme political and social control. Though reduced in both numbers and geographical mobility, by the beginning of the twentieth century there were signs of demographic recovery and population growth. However, it was not until after World War II that the Aboriginal peoples of Canada began to regain some political and social control of their lives. This coincided with a shift in government policy from the more traditional practices of assimilation towards policies of integration that have culminated in greater Aboriginal self-determination.

In view of these changing political circumstances for Canada's Aboriginal peoples since World War II, it is not surprising to find that their population growth rate has recovered, and now exceeds that for the total population. Indeed, much of the recent growth, especially that of registered Indians off-reserve, has been due directly to legislative change with reinstatements following the introduction of Bill C-31 (Norris 2001). Thus, the major growth between 1981 and 1991 of Aboriginal people residing outside Indian reserves was due primarily to ethnic mobility rather than residential mobility (Norris 2001).

Though still considerably less urbanized than the rest of the Canadian population, Aboriginal people are more mobile than the general population, but not substantially so. In 1996, just over half of the Aboriginal population were movers compared to just less than half of the rest of the population. More specifically, the majority of Aboriginal movers, as with non-Aboriginal movers, change residence locally. There is variability, however, in mobility between the various classifications of Aboriginal population suggesting the possibility of different impacts of, and responses to, government policy.

Those claiming multiple identities continue to experience the highest mobility rates, followed closely by Inuit, Métis and North American Indians.

The main exception to this general pattern is the very high local mobility rate for Inuit residing in non-census metropolitan areas at the time of the 1996 census. Most Inuit live in Nunavut in the remote Canadian north where the predominant form of mobility for each of the Aboriginal groups is local, and in every case exceeds that of the non-Aboriginal population. At the same time, the Inuit migration rate is relatively low. This would appear to reflect the increasing concentration of Inuit population in a few northern communities relatively isolated from each other, as well as the fact that these communities are encouraged through new self-governance structures and programmes to become more self-contained with respect to educational, social and health services.

The effect of demographic revival on Aboriginal mobility should also be recognized. The Canadian Aboriginal population is relatively young, and it is young adults in their twenties who display the highest mobility. Since this group constitutes a relatively large age cohort in the Aboriginal population compared to the rest of the population, and since Aboriginal fertility remains relatively high, it is expected that this youthful age profile will continue to inflate the overall Aboriginal mobility rate for the next few decades to levels in excess of that observed for the rest of the population.

In all of this, government policy and legislation has been, and continues to be, a major factor impacting on Aboriginal mobility and redistribution. Historically, the Indian Act of 1876, allowed the government of Canada to interfere in the lives Canada's Native Indians (Frideres 1998). For example, Indians were confined to reserves, and this ultimately served to lower their level of mobility to urban areas off-reserve. It also served to motivate many off-reserve Native Indians to move back to reserves in order to attain affordable housing which remains beyond reach for many in most urban centres. The Inuit were also subjected to direct influences of government policy as many were moved to new settlement areas when they would have preferred to remain in their original communities.

One consequence of assimilationist policies applied between 1950 and 1970, was the movement of many young people off their reserves to schools where they were exposed to the dominant culture in a pervasive manner. A measurable byproduct of this relocation was the loss of native language with fewer than 20 per cent of adult Aboriginal people currently able to speak their native tongue. While Aboriginal people must still leave reserves to receive university-level education, many now return after completion to participate in community development in line with current emphases in public policy.

Since the early 1980s, the Government of Canada has embarked upon a new policy approach to Aboriginal affairs. Abandoning its assimilationist goals, the government has pursued more integrationist policies that include the principles and practice of Aboriginal self-government. As a result, a devolution of government control has given way to Aboriginal control over an increasing number of fiscal areas, such as finance, education, and social services, that were hitherto solely within the purview of the federal government. As government continues to work towards granting Aboriginal people self-government with greater control over their destiny (for example, as it did with the establishment of Nunavut in 1999 for the Inuit), the movement of people away from Aboriginal communities or reserves is likely to be reduced. The actual rate will be dependent upon the extent to which Aboriginal institutions and employment opportunities are created for those who wish to remain in communities.

The Inuit are a case in point. Today, they are more involved in governing themselves and live in communities that can better provide for their social and economic needs. It also appears that by assuming a more sedentary way of life, Inuit have become more integrated into Canadian society than either the Native Indians or Métis. Perhaps similar integration of Native Indians and Métis will occur as they gain more economic and political independence and self-government. It is true that the major reasons for moving remain similar in many ways to those expressed by the non-Aboriginal population (Breton *et al.* 1990), with housing, family issues, and occupational and job opportunities among the most important. Nevertheless, many Aboriginal people still consider their communities and reserves as their homeland. As far as their future mobility is concerned the old adage "many go near, few go far" would still appear to be valid for Aboriginal people no matter what changes occur in their self-management and socioeconomic status in the twenty-first century.

References

Breton, R., Isajiw, W.W., Kalbach, W.E. and Reitz, J. (1990) *Ethnic Identity and Inequality*, Toronto: University of Toronto Press.

Clatworthy, S. (1996) *Migration and Mobility of Canada's Aboriginal Population*, Ottawa: Canada Mortgage and Housing Corporation.

Dickason, O.P. (1999) 'Aboriginals: Métis', in Paul R. Magocsi (ed.) *Encyclopedia of Canada's Peoples*, Toronto: University of Toronto Press.

Dorais, L. (1999) 'Aboriginals: Inuit', in Paul R. Magocsi (ed.) *Encyclopedia of Canada's Peoples*, Toronto: University of Toronto Press.

Frideres, J.S. (1998) *Native Peoples in Canada: Contemporary Conflicts* (4th edition), Scarborough: Prentice Hall.

Gabriel Dumont Institute (1994) *The Canadian Atlas of Aboriginal Settlements*, Regina: Gabriel Dumont Institute of Métis Studies and Applied Research Inc.

Gerber, L. 1984 'Community characteristics and out-migration from Canadian Indian Reserves: path analysis,' *Canadian Review of Sociology and Anthropology*, 22(2): 145–65.

Government of Canada (1999) *Basic Departmental Data*, Ottawa: Department of Indian Affairs and Northern Development.

Kydd, D. and Kredentser, S. (1992) *Land Use on Reserves, Surrenders and Designated Lands*, Vancouver: Legal Services Society of British Columbia.

Miller, J.R. (1999) 'Aboriginals: Introduction', in Paul R. Magocsi (ed.) *Encyclopedia of Canada's Peoples*, Toronto: University of Toronto Press.
National Film Board (1995) *Broken Promises: The High Arctic Relocation*, Ottawa: National Film Board of Canada.
Norris, M.J. (2001) 'Aboriginal mobility/migration and First Nation affiliation: the urban experience', Calgary: University of Calgary, Chair of Ethnic Studies Special Lecture Series.
Patterson, E.P. II (1972) *The Canadian Indian: A History Since 1500*, Toronto: Collier-Macmillan Canada.
Patterson, P. (1982) *Indian Peoples of Canada*, Toronto: Grolier.
Reeves, W. and Frideres, J.S. (1981) 'Government policy and Indian urbanization: the Alberta case', *Canadian Public Policy*, 7 (Autumn): 584–95.
Statistics Canada (1996) *Catalogue no. 92–351–XPE*, Ottawa: Statistics Canada.
Taylor, J. (1996) 'Surveying mobile populations: lost opportunity and future needs', in J.C. Altman and J. Taylor (eds) *The 1994 National Aboriginal and Torres Strait Islander Survey: Findings and Future Prospects*, Canberra: Centre for Aboriginal Economic Policy Research, The Australian National University.

Part II
Data issues and analysis

6 Data sources and issues for the analysis of Indigenous peoples' mobility

Bruce Newbold

> While the *fact* of frequent mobility among Indigenous people is acknowledged, the *facts* remain largely unknown.
>
> (Taylor 1996: 19)

Migration behaviour has long been of interest to economists, geographers, demographers and sociologists contributing to planning and policy formulation with respect to service provision and other issues of community interest. The economic and social implications, origin and destination effects and differences between migrants and stayers are often representative of other structures within a society and have been the focus of much of the migration literature over recent decades (Plane and Rogerson 1994). Yet, there is a conspicuous lack of information regarding the mobility behaviour of Indigenous populations. As recently as 1996, Taylor and Bell (1996a: 154) described the existing research on the migration of Indigenous peoples as "diffuse, partial and fragmented". This fragmentation can be traced to a variety of origins, not least of which are data deficiencies and the failure of standard mobility measures to define a population that may conceive of home as a regional concept rather than as a specific place.

The paucity of information on Indigenous movement is especially problematic when it is realized that the mobility of Indigenous populations is not just a subset of movements observed within the overall population. Evidence from Australia, where the study of Indigenous movement is among the best developed and articulated, shows substantial differences in mobility behaviour, patterns and impacts between Indigenous and non-Indigenous populations. The unique cultural and structural settings of Indigenous populations across New Zealand, Australia and North America, representing a common history of oppression, racism and discrimination have created the context for unique mobility outcomes. Moreover, the Indigenous population within any individual country is not a homogenous group. North American Indians, for example, differ socially, culturally and economically from Métis and Inuit in Canada, raising the possibility of further differences in migration behaviour. These are clearly evident in mobility differentials between groups and across space, expressed through rates and patterns (Norris 1990; Snipp 1989).

Most existing research on Indigenous mobility is based either on census data or on the results of ethnographic study. Compared to the general population, far less analysis is derived from survey data. The provision of ethnic identifiers in census data allows for consideration of Indigenous population movement in terms of the standard key perspectives on mobility, including movement propensities, the redistributive effects of migration, and the spatial patterning of flows (Bell 1995). Other perspectives, which focus more upon migration careers, have been less thoroughly explored for Indigenous populations, as is the case in migration research generally. Notwithstanding the widespread availability of Indigenous migration data from national census collections it should be noted that significant differences do occur between countries in terms of census timing, availability, geographic coverage and approaches to defining Indigenous populations. Such variation creates difficulty for multinational comparison of mobility and migration in general (Long 1988), but for Indigenous populations these problems are likely compounded by unique cultural factors, such as how different Indigenous groups interpret concepts such as usual residence, and the degree to which frequent, circular or short-distance moves associated with tradition, land and kin networks are a feature of mobility. Ultimately, Indigenous mobility and its conceptualization is very different from that of the general population.

These complex but complementary issues relating to methodology and culture in the study of Indigenous mobility demand further attention. Indeed, an understanding of the interplay between data sources, research methods and cultural context is indispensable to informed understanding of Indigenous mobility. This chapter endeavours to provide the essential background by reviewing data sources available for the analysis of Indigenous people's mobility in the four country settings, and examining the methodological problems associated with the use of such data.

Data sources

Spatial mobility encompasses a wide range of movement, including seasonal or temporary circulation, intra-urban residential mobility, longer distance moves involving a change of lifestyle or labour market area and international moves. *Migration* generally refers to those moves that involve changes in the labour market and lifestyle, while the term *mobility* is typically used to describe short-distance moves. For the purposes of this chapter, these terms will be used inter-changeably, as most of the issues discussed (except where noted) cut across these rather arbitrary distinctions.

The cornerstone of census-based mobility analysis is the standard definition of mobility which is based upon a change in usual place of residence, typically over a five-year period. If residences differ, a move is recorded, implying a permanent relocation. Using the concept of a single place of residence eliminates daily or seasonal migration, reflecting the view that migration should refer to a change lasting a significant period (Long 1988). While all censuses in Australasia and North America define the migration interval as five years, Canada (since 1991) and Australia (since 1976) have also collected data on usual place of residence one year prior to the census. This one-year interval provides additional insight into the size and nature of repetitive moves and chronic migration (Bell 1996; Newbold and Bell 2002). Most censuses also record migration status, which indicates whether an individual has changed residence or not, and capture some measure of distance moved such as "same house" (no migration), "different house, same county", etc. The Australian census records both *de facto* (place where individuals were actually enumerated) and *de jure* (places where usually resident) information, cross-tabulation of place of enumeration by usual place of residence, providing a crude, but useful insight into short-term migration (Taylor 1998).

Complete count census files are rarely available directly to users and special cross-tabulations that incorporate information on migration and mobility along with their other variables are often expensive to obtain. As an alternative, microdata files (with the individual as the unit of analysis) are available within the United States, Canada and Australia. These provide a flexible means of producing a wide variety of tabulations on selected characteristics and special populations, including information on mobility (Fotheringham and Pellegrini 1996).

In the United States, Public Use Microdata Sample (PUMS) files have been available from the decennial census since 1960, but have varied in content and definition over time (Census of Population and Housing 1992). Available as a weighted 5 per cent sample, the geography of the 1990 PUMS file utilizes Public Use Microdata Areas (PUMAs) as the base geography, while earlier (pre-1990) PUMS files used the county as the basic geographic unit. These PUMAs frequently follow county boundaries and have populations of approximately 100,000. However, large metropolitan areas are composed of several PUMAs while sparsely populated, adjacent counties are aggregated into one PUMA. Still, the geography of the PUMS files is superior to most comparable data sources, allowing the identification of intra- and inter-PUMA, intra- and inter-metropolitan, intra- and inter-state, and international moves. Another important feature of this data source is the ability to link to other census data, taking advantage of existing publications regarding the economic and social characteristics of an area (Fotheringham and Pellegrini 1996).

Statistics Canada has also provided Public Use Microdata Files (PUMFs) since 1971 (Statistics Canada 1999). Representing a weighted 2 (pre-1991 files) or 3 per cent (1991, 1996) sample of individuals (household and family files are also available), it is made available every five years and includes information on personal and household characteristics much like the United States' PUMS files. From a spatial perspective, the PUMFs are far less extensive than those available in the United States. Migration geography is limited to the ten provinces and two northern territories. Although home to a relatively large proportion of the Indigenous population, the territories are sparsely populated and are aggregated within the census files by Statistics Canada to protect confidentiality, confining migration analysis to an eleven-by-eleven matrix at the interprovincial scale. Frequently, however, small sample sizes in the four Atlantic provinces (Newfoundland, Nova Scotia, Prince Edward Island, and New Brunswick), the interior Prairie provinces (Manitoba and Saskatchewan) and the aggregated northern territories unit forces the consolidation of the migration matrix, commonly to a six-by-six matrix. This limited geography provides little insight into some of the more interesting questions associated with movement between metropolitan areas and/or rural and reserve areas among the Indigenous population since the usual place of residence five years ago is limited to the provincial or supra-provincial scale. Customized tabulations and extractions from Statistics Canada provide greater geographic detail and an additional option to the researcher. Utilizing a 20 per cent sample, they come at a cost to the user dependent upon the number of cells within the tables.

In contrast, researchers in Australia and New Zealand have been much more reliant upon customized extractions available from their statistical agencies. Both New Zealand and Australia conduct censuses every five years, with the most recent in both countries being the 2001 census. The New Zealand census distinguishes 1,774 Area Units within 16 Regional Councils, although it has only included a question relating to place of previous residence since 1971 (Carmichael 1993). Previously, inter-regional migration flows had to be estimated via survival and cohort analysis techniques (Pool 1991). The Australian census identifies 58 internal Statistical Divisions, based upon 1,336 Statistical Local Areas (SLAs) which are the lowest level for which usual place of residence data are available, and which can be aggregated to various scales. This allows, for instance, the analysis of intra-metropolitan mobility.

The Australian microdata are somewhat more restrictive. Notwithstanding the fact that the Australian Bureau of the Census has released a public use sample since the 1981 census, two main limitations constrain its use. First, the file represents only a 1 per cent sample of the Australian population. Although suitable for measuring general characteristics at the state or national level, the file is especially limited as a source of migration data on the Indigenous population given its relatively small size. In fact, this problem is generally applicable to other data sources such as the Canadian and United States microdata files, with sparse populations of the respective Indigenous groups limiting geographical definitions. Users must refer to estimation and reporting protocols within each sample in order to ensure unbiased results. Second, prior to the 1996 census, no data on previous place of residence one year ago were provided, although it was possible to distinguish those who had migrated over the five year period. Improvements in the 1996 unit record file meant the identification of 43 regions of residence one and five years earlier, as well as at the time of the census.

The variety of spatial and temporal boundaries used for measuring and comparing migration creates conceptual and empirical difficulties, as each one represents an arbitrary definition of space and time. The portion of the population that is defined as migrants depends not only on individual propensities to migrate, but also reflects the migration interval along with the size, shape and population distribution within the units being used (Long and Boertlin 1976; Newbold 1997; Plane and Rogerson 1994; Taylor 1992). In general, the larger the unit, the fewer moves are counted. What would be measured as an intra-provincial or intra-state migration in Ontario (Canada) or the Northern Territory (Australia), for example, would likely cross several state lines in the United States or regional council boundaries in New Zealand.

Although costly, customized extraction of information represents an alternative source of mobility data, with Statistics Canada, Statistics New Zealand, and the Australian Bureau of Statistics all performing this service at cost. Whether customized tabulations or "off the shelf" data such as microdata files are used, the usual caveats associated with the analysis of census based migration measures apply. Census-based measures exclude migrations that occur within the census interval, miss multiple and return migrations, and fail to capture an extended migration history (Bell 1996; Morrison 1971; Newbold 1997; Newbold and Liaw 1994; Plane and Rogerson 1994). Shorter migration intervals reduce the confounding effect of multiple moves, but are still subject to the other caveats. Census data further assume survival over the period and provide only a static picture of the dynamic process of population change (Taylor 1992).

Issues and concerns in the use of census data

In addition to these customary constraints on mobility research, further issues emerge when focusing on Indigenous populations, since most instruments and procedures are designed to capture the mainstream population. Conventional definitions of time and space are not always well-suited to capturing Indigenous movement and the omission of data on short-term, circulatory mobility is a serious deficiency. An additional concern is the effect of changing census definitions of Indigenous populations and changing propensities of individuals to identify as Indigenous when enumerated.

Methods of measurement: concepts of "home" and circulation

Ethnographic research suggests that circular mobility is very high in areas where Indigenous peoples form the majority of the population, involving daily and periodic movements for a wide range of purposes. From a geographical perspective, this is particularly true in sparsely populated areas, such as Canada's northern Territories and the northern and central parts of Australia. Despite this high mobility, very low rates of movement are recorded by the census among the Inuit in northern Canada (Norris 1990) and among Aborigines in Australia's Northern Territory (Taylor and Bell 1996b). Because the census utilizes fixed interval measures of movement, much significant Indigenous mobility goes unrecorded owing to its short-term or circulatory nature. Furthermore, movements among small, scattered populations may go unrecorded in regional analysis, while existing political and territorial divisions are unlikely to correspond to cultural and political divisions within Indigenous populations. This inability to reconcile census geography with the social geography of Indigenous groups can also obscure potentially significant differences between groups such as the Inuit and Native American Indians within Canada.

An additional problem underlying the use of census-based measures of mobility for some Indigenous populations is their reliance on unambiguously defined single points of residence. The difficulty here is that for cultural reasons many Indigenous people may perceive residential space as regionally defined rather than fixed in a single locality (Taylor 1996). This issue has been highlighted by ethnographic studies in Australia which have focused upon biographic and contextual analysis of population movements (Altman 1987; Bryant 1982; Cane and Stanley 1985; Martin and Taylor 1996; Taylor 1989; Young and Doohan 1989). These reveal a high propensity for short-term mobility among many Australian Aborigines, especially in sparsely populated areas where traditional cultural forms are intact. Many belong to extensive kin networks and may share residences for brief or extended periods, with circuits of movement linking kin over socially defined territories encompassing large geographic areas. Cyclical movements are also found to be associated with access to services (such as health care), marginal labour force attachment and seasonal employment opportunities. The socially porous nature of Indigenous communities and households may also contribute to high mobility and circulation (Finlayson 1991; Smith 1992). Drawing upon the Australian experience again, large interchange and movement within the Indigenous population has been observed between households with migrants relying upon kin and friends to meet short-term housing needs and for the socialization of children (Gale and Wundersitz 1982; Martin and Taylor 1996; Young and Doohan 1989). In one study of Aboriginal households in a remote north Australian town, up to 25 per cent of the resident population at any one time were estimated to be household visitors with a usual place of residence elsewhere (Taylor 1990).

In addition to ethnographic studies, the comparison of *de jure* and *de facto* place of residence census data allows some limited insight into short-term mobility. For example, at the 1991 census, as many as 7 per cent of

Aboriginal people in Australia were enumerated away from their usual place of residence, compared to the 4 per cent of the rest of the population (Taylor 1998). Most of these individuals were involved in local moves within the same region, although significant proportions undertook longer distance migrations especially from rural areas to urban central places. In some regional centres, for example, up to 12 per cent of the enumerated Aboriginal population reported their usual place of residence to be elsewhere.

The implication, for at least some Indigenous populations, is that the notion of a single usual place of residence is problematic. This has led some researchers to suggest that the concept of a "mobility region" may better define Indigenous perceptions of place and home with reference to a geographic area as "home", as opposed to geographic place (Taylor 1998; Uzzell 1976; Young 1990). Defining moves as geographic shifts beyond some minimum duration also remains arbitrary, and as noted this may conceal a good deal of significant Indigenous mobility. Following Roseman (1971), a more sophisticated approach would be to gauge a person's commitment to destination and origin areas to distinguish between "partial" and "total" displacement, although this necessarily implies alternative data sources and methodologies. While little such conceptualization is evident in the literature, it is likely that total displacement would be less prevalent among Indigenous populations given the relatively greater importance they are likely to attach to maintaining social, cultural and economic links with a multiplicity of places.

Ancestry, race and other issues

A basic problem pervading all the national data sets referred to here is the definition of Indigenous identity, with it becoming increasingly complicated to define for statistical purposes who is Indigenous, and who is not. With implications that include social, political and constitutional concerns, the boundaries of group membership are of central importance (Pool 1991; Snipp 1989, 1997). Theoretically, identifying the Indigenous should not be a problem, based upon self-identification with an Indigenous group and/or who is recognized by others as Indigenous (Eschbach 1993; Pool 1991; Snipp 1989). The reality is very different, with statistical and legal definitions frequently straying from this ideal over time, and varying between the United States, Canada, Australia and New Zealand, thus making it difficult to define Indigenous populations, to reconcile alternative data sources, to validate responses, and to make spatial and temporal comparisons.

The United States exemplifies the problems associated with various measures of identification and the evolution of society's broader view of the place of the Indigenous population relative to the mainstream. Definitions dating from the 1870s have included genetic and mystical bases, while bureaucratic definitions have included "blood quantum", with each race having a unique blood type which "determined" its physical appearance and social behaviour (Snipp 1989). Taxation definitions have also been applied: if Indians held United States Citizenship, they would be taxed and enumerated. Those residing on reservations were not taxed, did not hold citizenship and were not enumerated, except within special surveys.

Although recognizing the problems inherent within these definitions, the United States government has not attempted to provide a general definition of the Indigenous population, leaving it instead to the domain of tribal governments and bureaucrats. Prior to 1960, enumerators assigned race based on observation, clearly creating the possibility of reporting bias within the census. Censuses since (and including) 1960 have utilized the notion of self-identity. This has the advantages that no prior assumption is made about the concept of race, it does not require specific tests, and there is no explicit genetic theory or other attributes for group membership (Snipp 1989). But, it fails on two counts to provide an exclusive definition of identity. First, identity is open to personal interpretation, as each respondent uses their own calculus to define the concept. Second, recent United States definitions are based upon two competing definitions of race and ancestry (ethnic origin) within the census. Both concepts are defined by society and change over time. Census reporting of race reflects a choice among a list of races. Ethnic origin (ancestry) was added as a second identity item starting with the 1980 census and requires a written response. In both 1980 and 1990, the census bureau coded the first two ancestries listed, along with multiple ancestry combinations. In 2000, for the first time, reporting of two or more racial identities was also permitted (see Chapter 4).

Based upon calculations from the 1980 and 1990 PUMS, two divergent estimates of the North American Indian population are revealed (Table 6.1). Using race to define the Indigenous population results in an estimate of approximately 2.2 million (0.8 per cent of the total United States population) American Indians, including Aleut and Eskimo origins in 1990. Ethnic ancestry produces a very different figure, with a 1990 population of 5.2 million (2.0 per cent). The racial profile of all individuals reporting Indian ancestry is diverse, with 62.9 per cent reporting a white racial background, 2.9 per cent black, 30.6 per cent North American Indian and 3.0 per cent Asian backgrounds. A parallel analysis can be made of the distribution of ethnic ancestry among people reporting American Indian as their race (first ancestry only). Out of the 2.2 million people identifying their race as North American Indian, 73.9 per cent claimed North American Indian ancestry. Snipp (1989) reported a similar diversity of responses dependent upon the use of ethnic or racial definitions, although the number reporting American Indian ancestry was larger in 1980 than in 1990, which may be related to coding, methodological differences or self-identification issues. Interestingly, Eschbach (1993) concluded that the redistribution of the American Indian population was attributable

Table 6.1 Population estimates of the American Indian population by race and ancestry, 1980 and 1990

	1980[a]		1990[b]	
	Number	*%*	*Number*	*%*
Race				
American Indian	1,217,200	18.0	1,590,436	30.6
Black	256,700	3.8	149,073	2.9
White	5,173,500	76.6	3,270,797	62.9
Asian	31,800	0.5	155,807	3.0
Other	75,600	1.1	31,804	0.6
Total	6,754,800	100.0	5,197,917	100.0
Ancestry				
American Indian	1,126,760	73.2	1,590,436	73.9
African	3,040	0.2	2,946	0.1
European	161,640	10.5	193,064	9.0
Hispanic	50,820	3.3	77,295	3.6
Asian	4,980	0.3	11,677	0.5
Other	192,580	12.5	275,508	12.8
Total	1,539,820	100.0	2,150,926	100.0

Notes:
[a] Derived from Snipp (1989: 48–9).
[b] 1990 5% PUMS file.

to changes in identification rather than migration, reinforcing the need to clearly define the population at hand. Eschbach (1993) also noted that the 1980 PUMS files over-estimated the Indian population.

The diversity of responses led Snipp (1989: 51) to suggest three categories of American Indian identity. Those who define their race and ethnic background as American Indian could be considered as the "core" population. A second category, "American Indians of non-Indian ancestry", would include those who report their race as Indian but include non-Indian ancestry in their ethnic background. A third group includes those who cite a non-Indian race but claim Indian ancestry within their ethnic background, leading Snipp to call this group "Americans of Indian descent".

Although also relying upon self-identification, definition of Indigenous identity is no less complicated in other countries. Within Canada, definition of the Indigenous population typically includes Inuit, Métis, registered status Indians and non-status Indians, with differences across various censuses (Elliott and Foster 1995). Status definition requires particular attention. Dating from the Indian Act of 1874, Status Indian is a legal term referring to anyone whose name was on a band list (the basic organizational structure of Aboriginal peoples) or on the Indian registry. However, women lost their status if they married non-Indians (the same was not true for Indian men marrying non-Indian women), resulting in many individuals with native ancestry being non-status and vice-versa (Elliott and Foster

1995). It was not until 1985 that a process was established to reinstate status, although support for this legislation has been variable and inconsistent within the Indigenous community.

Notwithstanding the problem associated with the changing status identifier, identification of the Indigenous population through the Canadian census is also problematic, as definitions have varied from one census to another. The 1991 census, for instance, based Aboriginal origin on ethnic origin, which primarily measured ancestry. The 1996 census shifted the focus away from ancestral background to one based upon self-identification (first used in the 1991 Aboriginal Peoples Survey), so that the 1996 self-reported Aboriginal data are not comparable with either the 1991 or 1996 ethnic origin/ancestry values, as they measure two different concepts. For example, some that have Aboriginal ancestors do not see themselves as Aboriginal, and vice versa. Nonetheless, the 1996 census still allowed four separate, non-exclusive definitions of the Canadian Aboriginal population, referring to (i) ethnic origin (which includes Aboriginal origins), (ii) registered (status) Indian indicator, (iii) membership in an Indian band or First Nation, and (iv) an Aboriginal ethnic (ancestry) category. Based upon estimates drawn from the 1991 and 1996 Public Use Microdata Files (PUMF), Table 6.2 documents the various estimated populations. In 1996, 788,220 persons reported identifying with at least one Aboriginal group. However, the total Aboriginal identity population was 799,010, which includes 19,220 persons who did not identify with an Aboriginal origin group but who reported being a status Indian and/or who were members of an Indian band or First Nation (Statistics Canada 1999). Focusing only on the status Indian population in 1996 results in a count of 498,240 individuals, which is substantially lower than the Indian register count (610,900), with the difference associated with under-enumeration and some methodological and definitional differences between the two sources (Statistics Canada 1999). Unlike the United States, the Canadian census does not collect data on race and the number claiming a single Aboriginal ancestry is small (482,328). This probably reflects the inability to identify multiple ethnic origins (including Aboriginal origins) within the PUMF sample. In fact, Statistics Canada (1998) reports that over 1.1 million people reported Aboriginal ancestry in 1996. Comparison with the 1991 population estimates reveals the difficulty of comparisons over census

Table 6.2 Population estimates of the Canadian Aboriginal population, 1991 and 1996

	1991	*1996*
Status Indian	364,766	498,240
Self-identification	n.a.	788,220
Ethnic origin	450,233	482,328
Band membership	346,100	497,376

periods, with no self-identified population in 1991. Ethnic origin again misses those who claimed multiple origins. Differences in the ethnic origin questions between 1991 and 1996 further complicate the comparison.

Similar circumstances and problems are found in New Zealand. Here, the government has tended to assume two main cultural groups: Maori and Pakeha, the latter referring to those of European descent, although the non-Maori population includes people of Asian, Pacific Islander and other origin (Pool 1991). The Maori definition has tended to refer to New Zealand Maori exclusively, excluding Cook Island Maori who have been counted as non-Maori. Since 1971, measurement of Maori identity has been based upon changing conceptual bases, representing a shift from a biological to a subjective, self-identification criterion. This makes it difficult to compare (and measure) population size and changes to that population, although Gould (1992) offers some suggestions to overcome these difficulties. Prior to 1986, New Zealand utilized a "half or more blood" definition, which Pool (1991) argues was answered as if it were an ethnic identification question. Since then, Maori status within the census has been derived from self-identified ethnicity, although other reporting agencies within New Zealand have not consistently used this measure (Pool 1991). In both the 1991 and 1996 censuses, a second ancestry question was included, in part to correspond to the electoral act and to appease some Maori politicians. As in the United States, the use of ancestry to identify Maori background tends to inflate the number of Maori, with a significant proportion of those of Maori descent saying they are not Maori ethnically. For example, those who report Maori descent are then asked to identify their Maori tribe, with a significant minority unable to answer this question.

While the ethnic identification question tends to be more robust and socially meaningful than other measures, it is also problematic. First, individuals may record affiliation with more than one ethnic group, with Statistics New Zealand assigning priority to Maori backgrounds when people identify two or more ethnic origins. The prioritizing of Maori origins tends to inflate the Maori count while deflating the Pacific Islander and Pakeha populations. Second, the high degree of intermarriage between Maori and other groups confounds the ethnic identity measure, prompting some groups to claim that ethnic data may not be valid (Pool 1991). The apparent fluidity between the two main groups and their cultural convergence in terms of the adoption of traits and behaviour patterns further clouds the issue. Indeed, a portion of the population identifies equally with both groups, with individuals in smaller studies stating "no preference" for either group, although statistically they would be counted as Maori (Gould 1992; Pool 1991). Third, international movement also confounds statistical counts, with large numbers of Maori resident in Australia. Similarly, large movements between New Zealand and the Pacific Islands are observed owing to the coding of Maori and Pacific Islander ethnic responses as

Maori. Finally, there have been significant levels of "category jumping" between censuses, complicating temporal comparisons.

In contrast to Canada, the United States and New Zealand, the Australian census identifies Indigenous origins via a single question, based on self-identification. Although ancestry questions were included in the 1986 and 2001 censuses, they have rarely been used for analysis of the Indigenous population. Given the difficulties that have arisen elsewhere with multiple questions relating to ancestry and ethnic origin, this would suggest a more consistent outcome. Unfortunately, the Australian data have their own peculiarities with censuses from 1971 to 1996 allowing Indigenous Australians three options, identifying either as Aboriginal, Torres Strait Islander, or neither. In 1996, a fourth option was introduced, allowing identification as *both* an Aboriginal and as a Torres Strait Islander (Taylor 1997). The rationale for this option was to capture those who had not responded in previous censuses if they did not identify solely with one group or another (Taylor 1997: 2). Taylor (1997) reports that 10,016 people chose the new option, effectively claiming both origins. While this may provide a more accurate count, it complicates analysis because in order to establish the size of the individual populations, this value needs to be added to both the Aboriginal and the Torres Strait Islander groups, which effectively double-counts the population and changes the growth rates of both groups. The result, as Taylor (1997) shows, is that the Torres Strait Islander population grew either by 7 per cent between 1991 and 1996, or by 44 per cent, depending upon whether the dual-origin population is excluded or included.

Even without this double-counting, the Australian Indigenous population increased by 33 per cent between 1991 and 1996, which is about double the figure that could be accounted for by natural increase alone. The increase could be attributed to various sources, including a campaign by the Australian Bureau of Statistics to increase awareness of the census among Aboriginals and the importance of responding to it (Australian Bureau of Statistics 1996). In addition, increasing intermarriage between Indigenous and non-Indigenous partners has meant that the children of these couples must elect either Indigenous or non-Indigenous identities. There seems little doubt, however, that a rise in the propensity to identify as Aboriginal was a pre-eminent factor in that part of the rise not explained by natural population growth.

The Canadian Aboriginal Peoples Survey

Despite repeated calls for development of alternative data sources, the census remains the primary source of data on Indigenous mobility in all four countries. Alternative sources are limited and frequently derive from other investigations, in which mobility is rarely the primary focus, despite its importance with respect to the provision of services and other policy issues. Australia's 1994 National Aboriginal and Torres Strait Islander Survey (NATSIS) is an example of this. Although originally designed to include questions pertaining to mobility, these were largely dropped in the final stages of survey development, although some health-related mobility questions were retained (Taylor 1996). Ethnographic studies and small scale surveys more commonly incorporate a mobility dimension (see, for example, Young and Doohan (1989), Taylor (1989) and Bryant (1982)) but generally focus on localized settings which limits the scope for quantification leading to generalizable results. Longitudinal files, which would allow the detailed analysis of individual movements, are either non-existent or lack a sufficiently large Indigenous population or the requisite geographical detail.

Statistics Canada's 1991 Aboriginal Peoples Survey (APS) is a valuable exception to this dearth of Indigenous mobility data. The APS is a weighted national survey of persons aged 15 and over with Aboriginal identity (North American Indians, Inuit and Métis) living on-reserve, in settlements, and off-reserve. The public use file was released in 1995. The portion of the Canadian population that identifies with an Aboriginal origin and/or are status Indians were randomly selected from the total Aboriginal population, excluding the institutionalized. The survey was collected from two domains: communities with a high concentration of Aboriginal people (especially reserves and settlements), and the remainder of the province or territory.

The APS builds on the 1991 census and was designed, among other objectives, to determine the number, sequencing and duration of moves over a twelve month period prior to the survey (Statistics Canada 1995; Taylor 1996). Linkage to other determinants and types of mobility such as seasonal circulation to camps and health related migration can also be derived. While information on seasonal mobility is quite detailed, the file includes less than 3,000 who moved for health-related reasons, so there are the usual limitations associated with sample size. Further disaggregation by location or Aboriginal identity would quickly result in untenable sample sizes.

Mobility is represented through two dimensions, the first of which echoes the larger Canadian census being based place of usual residence five years previously. A second series of questions refer to moves within the 12 months prior to the survey, with moves being defined as changes of residence that last at least a month. In this way, the time and space constraints conventionally associated with census data are replaced by a more liberal interpretation of mobility which provides considerably greater detail on short-duration moves. Table 6.3 shows the mobility status of the Aboriginal population across these two dimensions. Overall, the results reveal consistently high mobility with North American Indians, Métis and the Inuit all exceeding the 45 per cent level recorded for the Canadian population as a whole over the five-year interval (Newbold and Bell 1999).[1]

Table 6.3 Mobility status of the Canadian Aboriginal population aged 15+ by type of move, 1986–1991 and 1990–1991

Mobility status	Number of movers			% of total population aged 15+		
	Indian	Métis	Inuit	Indian	Métis	Inuit
Five-year (previous place of residence)						
Total movers	150,467	45,975	11,663	53.0	57.6	57.3
Different CSD	90,059	27,433	9,057	31.7	34.4	44.5
Same CD	16,172	4,290	1,342	5.7	5.4	6.6
Different CD	31,694	8,719	608	11.2	10.9	2.0
Interprovincial	12,542	5,533	657	4.4	6.9	3.2
One-year						
Stayer	245,913	65,966	17,085			
1 move	30,116	10,176	2,911	10.6	12.8	14.3
2 moves	6,671	3,087	–	2.3	3.9	–
3+ moves	1,390	550	–	0.5	0.7[a]	–
Movers	38,177	13,813	3,287			
Total population	284,090	79,779	20,372			

Source: 1991 APS.

Note:
[a] High sampling variability.

Contrary to Norris (1990), the Inuit appear to be highly mobile over short distances, but less so over longer distances, with higher short-distance mobility rates and lower longer distance migration rates than either North American Indians or the Métis. The Inuit also display the highest propensity to move over a 12-month period.

Recognizing that many Aboriginals maintain links to traditional culture, the APS also recorded absences within the past 12 months. Table 6.4, which sets out the frequency of such moves, reveals some interesting effects. For example, approximately half (51 per cent) of the Inuit population spent an average of 5.9 weeks on the land in the past year engaged in traditional activities. Native American Indians and Métis were less likely to spend time on the land (19.5 and 13.5 per cent, respectively), although their average absence during the year was similar to the Inuit (6.0 and 5.0 weeks, respectively). As another proxy for mobility, a small but not insignificant proportion of each group resided in homes in two different communities over the past year, suggesting circulation between communities.

Finally, the APS enquired why respondents moved, with the top five reasons listed in Table 6.5. Of those reporting a reason, approximately 22 per cent of the Aboriginal population reported that the move was for family reasons. An additional 16 per cent indicated the move was associated with personal preferences, 13.3 per cent of all moves were to better housing (which has been a chronic problem for the Canadian

Table 6.4 Mobility proxies among the Canadian Aboriginal population aged 15+, 1991

Proxy	Indian	Métis	Inuit
2 homes	6.3	6.7	5.7
% on land	19.5	13.5	51.3
Average time spent on land	6.0	5.0	5.9

Source: 1991 APS.

Table 6.5 Top five self-reported reasons (%) for moving in past 12 months among the Canadian Aboriginal population aged 15+, 1991

Overall		Indian		Métis		Inuit	
Family	21.9	Family	22.1	Preference	21.9	Better	32.7
Preference	16.3	Preference	14.9	Family	20.3	Family	22.7
Better	13.3	Better	12.2	Better	12.4	Preference	12.3
Employment	9.6	New house	10.5	Employment	12.3	New house	7.4
New house	9.0	Employment	9.1	Location	9.2	Employment	5.3

Source: 1991 APS.

Aboriginal population), 9.6 per cent for employment reasons and 9.0 per cent to a new home. Although each group gave similar reasons, their importance varied, with the most notable difference being the nearly 33 per cent of the Inuit that indicated a housing-related move, with employment reasons ranking lower (5.3 per cent) relative to other groups.

Despite its strengths, the APS is limited in three important respects and consequently does not completely meet all of its objectives. First, a significant problem is the limitation of spatial information within the publicly released data file owing to confidentiality issues. The geography associated with mobility is particularly limited within the publicly released data file. Usual place of residence in 1991 is confined to major geographic region and the eight major metropolitan areas with the largest Indigenous populations. Sub-provincial geography distinguished those living on an Indian reserve/settlement (North American Indians only), within a Census Metropolitan Area (CMA),[2] other urban, and other rural, but previous place of residence was limited to identification of same province/region, another province/region and on- or off-reserve status (North American Indians only). Since no geographical information was provided on the previous province or CMA of residence, the APS file does not provide meaningful data on origin–destination flows, although overall rates, measures of circulation and the reasons underlying mobility can be ascertained.

A second limitation, reflecting distrust and on-going tensions between the Aboriginal population and the Canadian provincial and federal governments regarding land claims and federal versus tribal self-government, is that the APS under-represents some Aboriginal groups. Survey attrition was associated with groups refusing to participate in the 1991 census (excluding them from participation in the APS) while other communities would not participate in the APS. While the overall response rate to the survey was adequate, coverage errors were substantial (Statistics Canada 1995). Although the impact at the national scale would be small, it is likely to be more significant at lower geographic levels where Indigenous people form a higher proportion of the total population. Differences between on- and off-reserve populations are likely downplayed in the resulting sample (Newbold 1998).

Third, although Statistics Canada set out to establish the sequence and reasons for moves over the 12-month period, identification of the timing of these is limited within the public use file. While it is possible to identify those who made multiple moves in the period, the relative timing and reasons for each move are not available within the public files. Reasons for mobility within the 12 months prior to the survey are grouped in the sample file and only a single response is recorded, ignoring secondary reasons.

Conclusion

Despite long-term interest in migration issues, knowledge of Indigenous mobility is still limited and fragmented, partially reflecting the limitations of existing data sources. Microdata provide a potentially useful mechanism for understanding migration processes, but are frequently limited in terms of spatial disaggregation and population size. Moreover, such data still only provide a cross-sectional picture of the target populations with little information on circulation, inter-censal mobility and what may be a rather fluid concept of the "home" region within the Indigenous population.

Despite these issues, census data still provide the foundation on which to benchmark Indigenous mobility. This information can be used to ground further work, to inform new survey and policy tools, and to characterize Indigenous mobility patterns, facilitating verification and establishing the basis for further applied research. Enhancements are needed. In particular, data instruments need to consider alternative definitions of mobility through time and space in order to recognize circulation and local mobility patterns. Although the Canadian APS moved toward this goal, data release limitations constrain its success and it is still not widely used. Longitudinal data files also offer potential, but suffer traditional weaknesses with respect to geographical specificity. Unfortunately, data needs are greater than are likely to be fulfilled with new sources, forcing researchers to consider alternative methodologies.

Complicating future analyses of the Indigenous population is an increasing self-awareness and identification within Indigenous communities. While not a problem *per se*, this further complicates comparison and measures of population change over time, as this emerging population is likely to be much more diverse than in the past. Following Snipp (1989) and Gould

(1992), construction of a classification system based upon degree of ancestry or identification may prove optimal. For example, "core" populations are likely to be deeply attached to the Indigenous population, having knowledge of Indigenous languages, culture, and practices, while other groups will have some linkage via ancestry but little knowledge of their culture.

Notes

1 Mobility rates are not aged-adjusted to reflect differences in age structures between the overall and Aboriginal populations.
2 Statistics Canada defines several sub-provincial geographies. Census Metropolitan Areas (CMAs) refer to urban areas with populations greater than 100,000. Both Census Sub Divisions (CSDs) and Census Divisions (CDs) are statistical creations, with the CSD generally following municipal boundaries and CDs being intermediate in size between the CSD and the province, following county or regional municipality boundaries. It should also be noted that the Canadian Northwest Territories was officially split on 1 April 1999, creating the new Nunavut Territory in the eastern Arctic. Nunavut is predominantly populated by Inuit peoples (85 per cent). The remaining Northwest Territories (yet to be officially named) is populated by a (52 per cent) majority non-Aboriginal population, with the remaining population represented by Dene, Cree and other North American Indian, Inuit and Métis. Although Nunavut was not officially recognized in the 1991 census, it was included as a geographical location in the APS.

References

Altman, J.C. (1987) *Hunter-Gatherers Today: An Aboriginal Economy in North Australia*, Canberra: Australian Institute of Aboriginal Studies.

Australian Bureau of Statistics (1996) *Population Issues, Indigenous Australians, 1996*, ABS Catalogue No. 4708.0, Canberra: Australian Bureau of Statistics.

Bell, M. (1995) *Internal Migration in Australia 1986–1991: Overview Report*, Canberra: Australian Government Publishing Service.

Bell, M. (1996) 'Repeat and return migration', in P.W. Newton and M. Bell (eds) *Population Shift: Mobility and Change in Australia*, Canberra: Australian Government Publishing Service.

Bryant, J. (1982) 'The Robinvale community', in E.A. Young and E.K. Fisk (eds) *Town Populations*, Canberra: Development Studies Centre, Australian National University.

Cane, S. and Stanley, O. (1985) *Land Use and Resources in Desert Homelands*, Darwin: North Australia Research Unit, The Australian National University.

Carmichael, G.A. (1993) *Trans-Tasman Migration: Trends, Causes and Consequences*, Canberra: Australian Government Printing Service.

Census of Population and Housing (1992) Public Use Microdata Samples [machine-readable data files], US Bureau of the Census, Washington DC: Bureau of the Census.

Elliott, S.J. and Foster, L.T. (1995) 'Mind-body-place: A geography of Aboriginal health in British Columbia', in P.H. Stephensen and S.J. Elliott (eds) *A Persistent Spirit: Towards An Understanding of Aboriginal Health in British Columbia*, Vancouver: UBC Press.

Eschbach, K. (1993) 'Changing identification among American Indians and Alaska Natives', *Demography*, 30(4): 635–52.

Finlayson, J.D. (1991) 'Pacific island movement and socioeconomic change: metaphors of misunderstanding', *Population and Development Review*, 17(2): 263–92.

Fotheringham, A.S. and Pellegrini, P.A. (1996) 'Disaggregate migration modeling: a comparison of microdata sources from the US, UK and Canada', *Area*, 28: 347–57.

Gale, F. and Wundersitz, J. (1982) *Adelaide Aborigines: A Case Study of Urban Life 1966–1981*, Canberra: Development Studies Center, The Australian National University.

Gould, J.D. (1992) 'Maori in the population census, 1971–1991', *New Zealand Population Review*, 18(1&2): 35–67.

Long, L. (1988) *Migration and Residential Mobility in the United States*, New York: Russell Sage Foundation.

Long, L. and Boertlein, C. (1976) *The Geographic Mobility of Americans – An International Comparison*, Current Population Reports Series P-23, no. 64, Washington DC: US Department of Commerce, Bureau of the Census.

Martin, D.F. and Taylor, J. (1996) 'Ethnographic perspectives on the enumeration of Aboriginal people in remote Australia', *Journal of Australian Population Association*, 13(1): 17–31.

Morrison, P.A. (1971) 'Chronic movers and the future redistribution of population', *Demography*, 8: 171–84.

Newbold, K.B. (1997) 'Primary, return and onward migration in the U.S. and Canada: Is there a difference?', *Papers in Regional Science*, 76(2): 175–98.

Newbold, K.B. (1998) 'Problems in search of solutions: health and Canadian Aboriginals', *Journal of Community Health*, 23(1): 59–73.

Newbold, K.B. and Bell, M. (2002) 'Return and onwards migration in Canada and Australia: evidence from fixed interval data', *International Migration Review*, 35(4): 1157–84.

Newbold, K.B. and Liaw, K.L. (1994) 'Return and onward interprovincial migration through economic boom and bust in Canada, from 1976–81 to 1981–86', *Geographical Analysis*, 26: 228–45.

Norris, M.J. (1990) 'The demography of Aboriginal people in Canada', in S. Shiva, F. Trovato and L. Driedger (eds) *Ethnic Demography: Canadian Immigrant Racial and Cultural Variations*, Ottawa: Carleton University Press.

Plane, D. and Rogerson, P. (1994) *The Geographical Analysis of Population*, New York: Wiley.

Pool, I. (1991) *Te Iwi Maori*, Auckland: Auckland University Press.

Roseman, C.C. (1971) 'Migration as a spatial and temporal process', *Annals of the Association of American Geographers*, 61: 589–98.

Smith, D.E. (1992) 'The cultural appropriateness of existing survey questions and concepts', in J.C. Altman (ed.) *A National Survey of Indigenous Australians: Options and Implications*, Research Monograph No. 3, Canberra: Center for Aboriginal Economic Policy Research, The Australian National University.

Snipp, C.M. (1989) *American Indians: The First of this Land*, New York: Russell Sage Foundation.

Snipp, M.C. (1997). 'Some observations about racial boundaries and the experiences of American Indians', *Ethnic and Racial Studies*, 20(4), 667–89.

Statistics Canada (1995) *The 1991 Aboriginal Peoples Survey Microdata File: Adults User's Guide*, Ottawa: Statistics Canada.

Statistics Canada (1998) *The Daily*, 13 January 1998, Ottawa: Statistics Canada.

Statistics Canada (1999) *1996 PUMF on Individuals, User Documentation*, Ottawa: Statistics Canada.

Taylor, J. (1989) 'Public policy and Aboriginal population mobility: insights from the Northern Territory', *Australian Geographer*, 20: 47–53.

Taylor, J. (1990) 'The estimation of small town Aboriginal population change', *Journal of the Australian Population Association*, 7(1): 40–56.

Taylor, J. (1992) 'Population mobility: policy relevance, survey methods and recommendations for data collection', in J.C. Altman (ed.) *A National Survey of Indigenous Australians: Options and Implications*, Canberra: The Australian National University.

Taylor, J. (1996) 'Surveying mobile populations: lost opportunity and future needs', in J.C. Altman and J. Taylor (eds) *The 1994 National Aboriginal and Torres Strait Islander Survey: Findings and Future Prospects*, Canberra: The Australian National University.

Taylor, J. (1997) 'Changing numbers, changing needs? a preliminary assessment of Indigenous population growth 1991–96', *CAEPR Discussion Paper No. 143*, Canberra: Centre for Aboriginal Economic Policy Research, The Australian National University.

Taylor, J. (1998) 'Measuring short-term population mobility among Indigenous Australians: options and implications', *Australian Geographer*, 29(1): 125–37.

Taylor, J. and Bell, M. (1996a) 'Population mobility and Indigenous peoples: the view from Australia', *International Journal of Population Geography*, 2(2): 153–69.

Taylor, J. and Bell, M. (1996b) 'Mobility among Indigenous Australians', in P.W. Newton and M. Bell (eds) *Population Shift: Mobility and Change in Australia*, Canberra: Australian Government Publishing Service.

Uzzell, D. (1976) 'Ethnography of migration: breaking out of the bi-polar myth', in D. Buillet and D. Uzzell (eds) *New Approaches to the Study of Migration*, Houston, TX: Rice University.

Young, E.A. (1990) 'Aboriginal population mobility and service provisions: a framework for analysis', in B. Meehan and N. White (eds) *Hunter-Gatherer Demography: Past and Present*, Oceania Monograph 39, Sydney: University of Sydney.

Young, E.A. and Doohan, K. (1989) *Mobility for Survival: A Process Analysis of Aboriginal Population Movement in Central Australia*, Darwin: North Australia Research Unit, The Australian National University.

7 Registered Indian mobility and migration in Canada

Patterns and implications

Mary Jane Norris, Martin Cooke, Daniel Beavon, Eric Guimond and Stewart Clatworthy

The Aboriginal population in Canadian cities has grown substantially since the 1960s. This growth has occurred in terms of absolute numbers, as a proportion of the total Aboriginal population, and as a proportion of the total urban population. The growth of the urban Aboriginal population has been commonly characterized as resulting from large numbers of people moving from Indian reserves to cities, searching for employment and relief from crushing poverty in their home communities. Two recent newspaper articles provide populist examples of this characterization:

> Manitoba's reserves are pockets of desperate poverty . . . islands of poverty can emerge when a large number of people reject the option of going where the jobs are even though their home community offers little in way of economic opportunity . . . Rural "Aboriginal" chiefs must accept that there are times when the only way to escape poverty on the reserve is to leave.
>
> ("Escaping Poverty", *Winnipeg Free Press*, 8 April, 1999)

> Some Aboriginal communities are lucky enough to sit on oil and gas deposits; many of them, however, have few natural resources. The reserves are clearly inadequate as generators of economic activity. *Hence the exodus to urban areas* [emphasis added]. The reserves are caught in a country-wide shift from rural to urban Canada. People are moving from rural to urban Canada, and from northern areas to southern cities.
>
> (Jeffrey Simpson, *Globe and Mail*, 19 May, 1999)

Both of these quotes, while examples of commonly held beliefs, reflect a serious misunderstanding of the actual patterns of Aboriginal mobility in Canada. This is especially so in relation to the migration of Registered Indians (those persons who are legally registered as an Indian according to Canada's Indian Act), the majority of whom live on-reserve. As this chapter will demonstrate, there has been no mass exodus among Registered Indians to urban areas, and the growth that has occurred in the urban

Registered Indian population has been much more the result of legislative change and natural increase than of migration. As well, it does not appear that all migration of Registered Indians is necessarily undertaken in search of employment, despite the way it is often characterized. An analysis of census data reveals high levels of mobility due largely to movement off-reserve by Registered Indians combined with considerable return migration to reserves. Rather than an exodus from reserves into cities being the main issue of policy concern, it is the overall frequency of movement leading to residential instability that has the greatest consequences for the well-being of Aboriginal people and the communities in which they live.

This chapter has two broad aims. First, it uses census data to describe and quantify recent patterns of mobility and migration among the Registered Indian population with particular focus on interactions between reserves and metropolitan centres. Second, it considers reasons for migration and residential "churning" of the population as a prelude to examining some of the consequences for policy and effective service delivery. [1] This concept of "churn" or "turbulence" is borrowed from analyses of mobility in the context of the developing world, in which the pattern often involves movement between rural and urban areas (Chapman 1978).

Distribution of Aboriginal populations

The term "Aboriginal" refers to the descendants of the original inhabitants of Canada. Canada's 1982 Constitution Act recognizes three broad cultural groups of Aboriginal peoples, including North American Indian, Métis, and Inuit, without providing legal criteria for membership. A further distinction is made by Canada's Indian Act which provides a legal definition of a particular North American Indian population by establishing the criteria for a person to be recognized as a Registered Indian in the Indian Register administered by Indian and Northern Affairs Canada (also known as the Department of Indian Affairs and Northern Development). On this basis the Aboriginal population can be classified into four major groups: Registered (or "status") Indians, who are registered under the Indian Act; non-status Indians, who have lost or never had status under the Indian Act;[2] Métis,[3] who are of mixed Indian and non-Indian origins; and Inuit, who reside mainly in two of Canada's three Arctic territories (the Northwest Territories and Nunavut) and in northern Quebec and Labrador. In the case of both status and non-status Indians, the term "First Nations" refers to the Indian people of Canada, regardless of status.

The concept of Aboriginality is multi-dimensional, such that the concepts of origins and identity, as well as legal definitions, have been variously used in the census of Canada to establish the size and composition of Canada's Aboriginal populations. In the 1996 census, there were 1.1 million people with Aboriginal ancestry and 799,000 reporting that they identified themselves as members of an Aboriginal group, or were registered under the Indian Act. Of these, the Registered Indian population (488,100) represents the largest of the four groups, followed by Métis (210,000), non-status Indians (90,400), and Inuit (41,100).[4]

Among Aboriginal populations, the distinction between status and non-status is an important one for any demographic analysis. There are certain rights and benefits associated with Registered Indian status, especially on reserves, where the majority of Registered Indians are located. These benefits include such things as access to funding for housing, post-secondary schooling, and tax exemption status, as well as land and treaty rights. Aboriginal populations in other communities, such as those of the Métis and the Inuit, do not have legal access to the same rights and benefits, as do Registered Indians on reserves. For these reasons, the distinction between reserve and non-reserve geographies is important in terms of understanding the push–pull factors associated with the migration patterns of Registered Indians. As discussed later, an understanding of the high mobility between reserves and cities of the Registered Indian population can be developed through comparison with the mobility of other Aboriginal groups, and the dynamics of migration between non-reserve Aboriginal communities and cities. This varied landscape of rights and needs of Aboriginal communities and urban populations is an important consideration from both government and First Nation perspectives in assessing the factors and implications associated with migration, such as programme and service delivery both on- and off-reserve.

Four categories of place of residence are used in analysing the migration patterns of Aboriginal populations. They include: reserves and settlements; urban Census Metropolitan Areas (CMAs); urban non-CMA; and rural areas. An *Indian Reserve* is legally defined in the Indian Act as a tract of land that has been set aside for the use and benefit of an Indian band or First Nation. *Settlements* include Crown land and other communities with Aboriginal populations as defined by Indian and Northern Affairs Canada, but do not include all Métis and Inuit communities. Reserves and settlements are combined as one geography for origin–destination flows. A *CMA* is a very large urban area, including urban and rural fringes, with an urban core population of at least 100,000 and, for this analysis, excludes rural fringes and any reserves in urban areas. An *urban non-CMA* is a smaller urban centre not designated as a CMA. It includes cities in which the urban core is at least 10,000, and for this analysis, excludes rural fringes and any reserves in urban areas. *Rural* areas comprise sparsely populated lands lying outside urban areas, including the rural fringes of CMAs, but excluding reserves and settlements.

For the purpose of analysis, movers between these places of residence are classified into two types, "migrants" being those who have moved between communities, and "residential movers" referring to those who have moved between residences in the same community. The term "mobility" applies to all moves involving a change of residence.

Patterns of Registered Indian migration

The first suggestion of a substantial flow of Registered Indians into cities from reserves was made in the 1960s with a search for better employment opportunities advanced as the primary cause (Hawthorn 1966). It was predicted that many reserve communities would eventually cease to exist because they experienced more out-migration than their numbers and replacement could support. The first conclusive presentation of quantitative data on the migration of Aboriginal people was provided by Siggner (1977), using 1971 census data. This analysis showed that the destination for 28 per cent of Registered Indians who had moved between 1966 and 1971 was a Metropolitan Area, while 27 per cent had an on-reserve residence as their destination. The strongest single out-migration flow was from rural, non-reserve areas, with people moving from these into both reserves and urban areas. Thus, rather than there being a unidirectional flow from reserves to cities, both reserve communities and urban areas were net gainers of migrants between 1961 and 1971.

This pattern of positive net migration to both reserves and cities continued throughout the 1970s and 1980s, indeed the volume of population exchange between reserves and urban areas intensified during this period with migrants to reserves from urban areas outnumbering those leaving reserves (M.J. Norris 1992). Nearly two-thirds of the net gain of 7,500 migrants by reserves between 1971 and 1981 was due to the flow from urban areas. Once again, 1991 census data reveal essentially the same pattern with both reserves and CMAs experiencing net migration gain, although the gain by reserves was greatest (Clatworthy *et al.* 1997: 31). Both rural non-reserve areas and smaller urban areas experienced a continued net loss of Registered Indian migrants between 1986 and 1991.

Figure 7.1 shows the net migration for each of the four geographies, for the 1966 to 1996 census periods. The pattern of movement is somewhat confounded by general North American migration trends, and particularly by the "metropolitan turnaround" that was experienced by the North American population in the later 1970s and early 1980s. On the whole, however, the census-based evidence does not support the idea of large-scale net migration gains of Registered Indians from reserves into cities. It does, however, reveal that the Registered Indian population has been highly mobile, and consistently more so than the Canadian population as a whole. This is especially so for those who do not live on reserves, or in other Aboriginal communities or settlements (M.J. Norris 1985, 1990, 1996), such that Registered Indian residents of cities are more mobile than are non-Aboriginal residents. Once again, this is a long-standing observation as analysis of 1981 census data also showed that Inuit and Aboriginal populations living in the south of the country were more mobile than their counterparts living in northern areas, and were also more mobile than non-Aboriginals in the south (D.A. Norris and Pryor 1984; Robitaille and

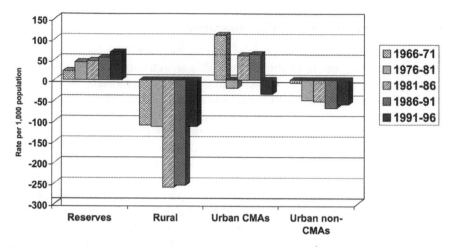

Figure 7.1 Five-year net migration rates for Registered Indians, 1966–1996.
Source: Norris 2000; Statistics Canada 1971, 1981, 1986, 1991 and 1996 censuses of Canada.

Choinière 1985). Finally, Clatworthy (1996) has established higher mobility among the Aboriginal identity population off-reserve, with more than 70 per cent of the urban population having moved over a 5-year period.

Migration as a component of population growth

The fact that *both* reserves and larger urban areas have been consistent net gainers of migrants undermines popular notions of a continuing exodus of people from reserves. No doubt such misperceptions arise due to the growth that has occurred in the Aboriginal population in urban areas, in particular among Registered Indians. However, as Clatworthy *et al.* (1997) note, there are several factors in addition to migration that influence the demographic growth of the Registered Indian population: fertility, mortality, family formation, intermarriage, legislative reinstatements, Indian registration, and band membership rules[5]. Registered Indian Status is legally defined, and the size of this population is therefore subject to variation as a result of changes in legal definitions. As well, census data rely on individual self-reporting of ethnic identity and there is evidence that, as individual awareness and ethnic self-identity changes, so too can the reporting of these affiliations in responses to the census questionnaire. This phenomenon is known as "ethnic mobility" and can itself be considered a component of growth in determining the size of different Aboriginal populations, as demonstrated by Guimond (1999).[6] Any one of these factors could result in differential growth rates between reserves and cities, although close examination shows that over the past decade the largest part of the increase in the Registered Indian population off-reserve has been due to legislative reinstatements resulting from changes to the Indian Act in 1985 (commonly referred to as Bill C-31).

To date, there have been over 120,000 persons who have been reinstated as Registered Indians. Data from the Indian Register show that as of the end of 1995, only 15.6 per cent of the Bill C-31 registrants were residing on-reserve (Clatworthy *et al.* 1997: 16). The fact that almost 85 per cent of the Bill C-31 registrants continue to reside off-reserve has meant that from 1985 to 1995 there has been a dramatic shift in the on-off-reserve population split. While the Registered Indian population living off-reserve grew over this period from 201,090 to 260,755,[7] an estimated 42 per cent of this growth was due to Bill C-31 additions. Against this background, it is worth noting that the census counts of the on-reserve population also increased significantly between 1991 and 1996 (from 184,710 to 227,285),[8] further substantiating the fact that reserves are not experiencing a mass exodus of their population.

Migration patterns 1991 to 1996

While the census provides the most complete picture of the patterns and trends of migration in Canada, there are several cautionary notes that one should consider with respect to the use of census data to measure the migration and mobility patterns of Registered Indians. First, census questions on ethnicity and mobility are only administered to a sample of the total census population. This does not include persons in institutions such as prisons, chronic care facilities, or rooming homes. This "missed" population could be problematic in that the incarceration rates for Aboriginal people are extremely high, particularly in the western provinces. Also, because of lower rent costs, there are very high concentrations of Aboriginal people living in rooming houses in urban centres. Under-reporting may also differ for males and females, and therefore could affect observed differences in the migration patterns of men and women. For example, a higher under-reporting of adult males because of incarceration may affect the observed gender differentials in migration.

Second, a significant proportion of the reserve population is not captured in the census due to the incomplete enumeration of certain reserves, as well as high rates of undercoverage on reserve. Since 1986, many First Nation communities have refused to participate in the census for a variety of reasons, such as an expression of their sovereignty, or distrust of government. Incompletely enumerated reserves often make trends over different census years more difficult to interpret because it is not always the same reserves that are participating from census to census, and also because of the impacts of differential undercoverage between censuses, both on and off reserves. Caution is also required in the comparison of migration data over time, because the population of Registered Indians is not directly comparable over census periods in terms of concepts and measurement. To some extent, reserves may be understated as a destination due to the fact that incompletely enumerated reserves are not represented in the current destination data, although they are in the origin as place of residence five years ago. For purposes of analysing the census data on migration flows incompletely enumerated reserves were excluded as origins in their respective censuses.

Finally, geographic units sometimes change over time. For example, new reserves may be created from census year to census year, or the geographic designation of a city as a CMA or non-CMA can change. Thus, some caution must again be used in making comparisons among geographic areas such as census metropolitan areas over time. Notwithstanding these caveats, clear patterns of spatial interaction may be discerned.

Net migration flows

As indicated, an important aspect of the migration patterns of Registered Indians that distinguishes them from other Aboriginal groups is their movement to and from reserves, especially between reserves and cities. Between 1991 and 1996, 61 per cent of out-migrants from reserves moved to urban areas (CMA and non-CMA), while 69 per cent of in-migrants to reserves came from urban areas. This continues the pattern that has been seen for the past five census periods: that both reserves and cities have been the major destinations. Regardless of origin (from reserves or other communities) large cities or urban CMA were the major destinations for 29 per cent of Registered Indian migrants in 1996, followed by 28 per cent each for reserves and smaller cities, with the remaining 15 per cent of migrants moving to non-reserve rural areas.

While there is some attraction to urban areas, the stream of migration from reserves to cities is smaller than the flow from cities to reserves. Overall, seven out of ten Registered Indian migrants over the 1991–1996 period can be classified into one of three major flows: urban-to-urban (37 per cent), urban-to-reserve (20 per cent) and rural-to-urban (13 per cent). Flows from reserves to urban areas (CMA and non-CMA) accounted for only 7 per cent of the migration volume. Between 1991 and 1996, for every 1,000 Registered Indians on reserve, only 38 had migrated out over the five-year period[9] compared to much higher out-migration rates of Registered Indians from small cities (non-CMA; 258 per 1,000), rural communities (288 per 1,000) and large urban areas (CMA; 192 per 1,000).

The pattern of net migration between reserves, rural, and urban areas over the 1991–1996 period, as shown in Figure 7.2 from the 1996 census, was largely similar to that observed in previous censuses in terms of five-year net migration flows. The overall effect of Registered Indian migration patterns for the 1991–1996 period is a net inflow to reserves of about 14,100 migrants and a net outflow of 6,400 migrants from rural areas,

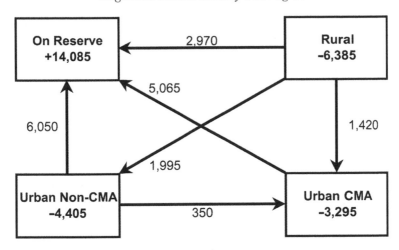

Figure 7.2 Five-year net migration flows for Registered Indians aged 5+, Canada, 1991–1996.

Source: Norris *et al.* 2001: 24. Reproduced with the permission of the Minister of Indian and Northern Affairs, 2002.

3,300 from large cities (urban CMAs), and 4,400 from the smaller cities (urban non-CMAs). Migration data for the one-year period 1995–1996 reflect this same pattern of loss and gain.

From this analysis of flows it would seem that while the major focal points in Registered Indian migration continue to be urban areas and reserves, the impact in terms of net gain or loss of population is felt most significantly in rural areas, which have lost Registered Indian population through migration mainly to urban areas. On the other hand, large inflows to urban areas are overshadowed by larger outflows of urban population to reserves. Thus, migration is a reciprocal process. Reciprocal moves between on- and off-reserve locations accounted for about a third of the 87,400 Registered Indians who were recorded as migrants over the 1991–1996 period, while only 3 per cent moved between reserves. As much as 64 per cent of migrants were involved in exchanges between off-reserve locations. Nearly two-thirds of the migration between on- and off-reserve locations involved migration from urban areas to reserves, and well over half of the migration between off-reserve areas was between urban areas.

Age-sex patterns of mobility and migration

Analysis of mobility and migration rates by age and sex provides insight into the process and role of migration across the life cycle of individuals. Mobility is associated with education, labour force transitions (employment, job loss, retirement), household and family formation and dissolution (marriage, divorce, widowhood). These events and life-cycle stages are just

144   *Mary Jane Norris* et al.

as age- and sex-dependent among Registered Indians as they are for the population as a whole. Accordingly, mobility rates among Registered Indians have consistently followed the standard age pattern for all Canadians: low over the school-age years, peaking during the young adult years of 20–29, and then declining fairly steadily thereafter, with young women in the 15–29 age group displaying higher mobility than their male counterparts (M.J. Norris 1985, 1990, 1996). This pattern has continued to the present. Thus, between 1991 and 1996, among the Registered Indian population aged 20–24, 829 per 1,000 Registered Indian women living off-reserve had moved, compared with 723 for males (Figure 7.3). Similarly, for Canadians, the female rate was higher than that for males (617 versus 511). Some of this gender difference among youth and young adults is attributable to younger ages at marriage and earlier entry into the labour force of females, events that are frequently associated with geographical movement.

In general, Aboriginal people who live outside their native communities and settlements have been found to move to a greater extent than both those living in Aboriginal communities and the general population (M.J. Norris 1985, 1990, 1996). This phenomenon continues to be evident, particularly among young people. For example, between 1991 and 1996, for every 1,000 Registered Indian women aged 20–24 living off-reserve, about 829 had moved at some point over the five-year period, compared with about 554 per 1,000 among those living on-reserve, and 617 for Canadians in general (Figure 7.3). Overall, mobility and migration rates of

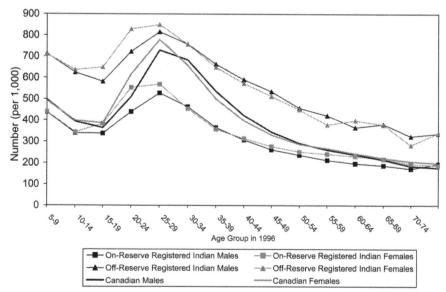

Figure 7.3 Number of movers per 1,000 population by age and sex: Canada, 1991–1996.

Source: Norris *et al.* 2001: 30. Reproduced with the permission of the Minister of Indian and Northern Affairs, 2002.

Registered Indians off-reserve (655 and 286 per 1,000 respectively) are much greater at all ages than the corresponding on-reserve rates (381 and 123) and higher than rates for the Canadian population in general (430 and 202). The higher mobility and migration rates of the off-reserve Aboriginal population is only partly attributable to movement from reserves and settlements since they also reflect residential movement within the same community as well as migratory movement to and from different communities.

The contrast in mobility between the Registered Indian population off-reserve and the total Canadian population is more pronounced among residential movers than among migrants, as shown by Figures 7.4 and 7.5. Conversely, the contrast between the on-reserve population and all Canadians is greatest among migrants. These comparisons demonstrate that while Registered Indians off-reserve tend to migrate, or change communities more than the average Canadian, they also change dwellings within a community or city more than do Canadians in general. On the other hand, the residential mobility of Registered Indians living on reserves is more similar to that of other Canadians. This is to say that across all age groups, people in reserve communities change their local place of residence at about the same rate as other Canadians.

Census data have consistently shown that women have higher rates of out-migration from reserves, and this is again true for the 1991–1996 period, especially among youth aged 20–24 (Figure 7.6). However, migration

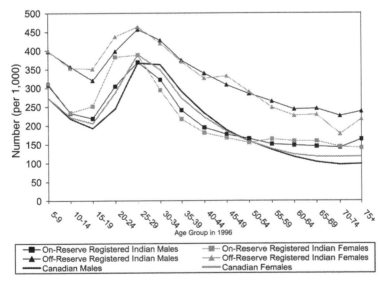

Figure 7.4 Number of residential movers per 1,000 population by age and sex: Canada, 1991–1996.

Source: Norris *et al.* 2001: 31. Reproduced with the permission of the Minister of Indian and Northern Affairs, 2002.

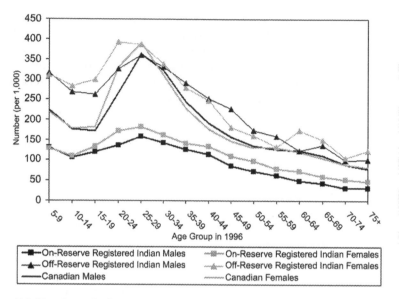

Figure 7.5 Number of migrant movers per 1,000 population by age and sex: Canada, 1991–1996.

Source: Norris *et al.* 2001: 32. Reproduced with the permission of the Minister of Indian and Northern Affairs, 2002.

from off-reserve locations to reserves contrasts sharply with migration from reserves, particularly with respect to gender differentials as men migrate to reserves at a greater rate than women do, especially from age 25–29 onwards. As will be discussed later, these gender differentials in the propensity to migrate to and from reserves suggest that men and women experience different push–pull factors in their decisions to migrate. The rate of migration for both males and females from locations off-reserve to reserves is higher than their out-migration from reserves, according to both one-year and five-year data. For example, for males between 1991 and 1996, the rate of in-migration to reserves was three times the out-migration from reserves (100 out-migrants per 1,000 off-reserve population, compared to 33 per 1,000 on reserve). The contrast in rates between the two flows is less pronounced for females (83 out-migrants per 1,000 off-reserve population compared to 43 per 1,000 on reserve).

Compared with the movement to and from reserves, males and females differ less in their migration between communities off-reserve for both the one-year and five-year data. The data show that among youth (15–24) women have slightly higher rates of migration than men, a pattern that is generally consistent with most migration streams. Both males and females have significantly higher rates of migration between off-reserve locations, compared to their rate of movement to or from reserves. Overall, for every

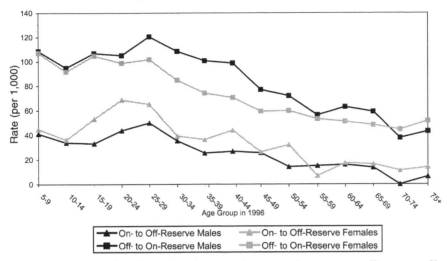

Figure 7.6 Registered Indian migration rates by age and sex: on- to off-reserve, off-
to on-reserve, and between off-reserve locations: Canada, 1991–1996.

Source: Norris *et al.* 2001: 36–7. Reproduced with the permission of the Minister of Indian
and Northern Affairs, 2002.

1,000 Registered Indian males living off-reserve, 236 had migrated between
off-reserve locations over the 1991–1996 period, compared to a rate of 100
per 1,000 moving from off-reserve to on-reserve and an out-migration rate
of 33 per 1,000 moving from reserves.

Community migration patterns

While examination of age- and sex-specific migration and mobility rates
provides important information for understanding the characteristics of
migrants in relation to life-cycle events, it is equally important to examine
the rates of in- and out-migration for different types of communities. This
is especially the case because of the role played by reserves in the overall
origin–destination patterns of migration. Between 1991 and 1996, First
Nation reserves or communities experienced higher rates of in-migration
than out-migration, such that there was a net inflow to reserves. This
pattern contrasts with that found in Inuit and Métis communities, the
members of which are generally not registered under the Indian Act. Those
communities experienced little difference between the rates of in- and out-
migration of non-registered Aboriginal people, with the result that there
was practically no overall population gain or loss due to migration. As
noted earlier, the persistent pattern of net inflows to reserves since the
1960s suggests the role of the benefits, available to Registered Indians
living on reserves, as a pull factor. Migration rates for the larger First
Nation reserves show that six of the top ten (according to population size)
experienced marginal net inflows (Figure 7.7).

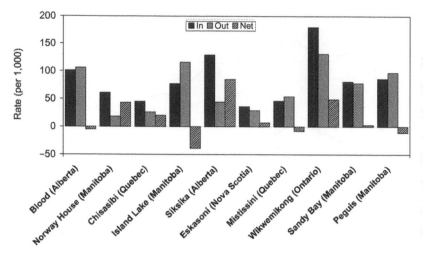

Figure 7.7 Five-year migration rates for most populated First Nation communities, 1991–1996.

Source: Norris *et al.* 2001: 41. Reproduced with the permission of the Minister of Indian and Northern Affairs, 2002.

However, the communities that make up the Island Lake Tribal Council in Manitoba, experienced a moderate net outflow, while Siksika in Alberta posted a relatively high net inflow. These cases demonstrate that despite overall trends in migration, the situation of each community is a unique response to prevailing influences. Important factors here can include distance to urban centres and available modes of travel, special needs of the community, and the availability of services.

For example, in the case of Island Lake, which includes four communities on a relatively remote northern reserve in Manitoba with no road access, the net outflow of its population may be linked with the unusually high prevalence of diabetes in those communities (Young *et al.* 2000). Because of special needs for dialysis machines many of the communities' members and their families must either fly or relocate to Winnipeg, the closest major service centre. This is consistent with 1996 census origin–destination data which show unusually high rates of in-migration to Winnipeg from Island Lake, particularly given its considerable distance from the city. On the other hand, Siksika is a relatively prosperous large reserve in Alberta, with well-developed business and governance structures, within commuting distance from Calgary (cf. Howes 2000). Furthermore, the community's innovative on-reserve housing encourages home ownership through financing as an alternative to community-provided housing (Melting Tallow 2001).

Analysis of five-year rates of in- and out-migration to and from large Census Metropolitan Areas (CMAs) also supports the observation of higher mobility and migration of Registered Indians in comparison to other Aboriginal groups and other Canadians. Figure 7.8 shows the top ten CMAs based on Registered Indian population and reveals that Registered Indians consistently have the highest rates of both in- and out- (and consequently gross-) migration, followed by other Aboriginal groups[10] and then by Canadians in general. Over the 1991–1996 period, most of these top ten CMAs generally recorded net losses of population through migration for all three groups, with the notable exceptions of Saskatoon and Thunder Bay which experienced net gains of Registered Indian and "other Aboriginal" migrants.

As can be seen from Figure 7.8, there is clear variation between CMAs in terms of their in-, out-, and net migration rates. It is likely that this reflects variable push and pull factors, including the presence in some cases of sizeable urban Aboriginal populations (such as in the western cities of Regina and Winnipeg), or educational institutions (such as exist in Thunder Bay). However, without further analysis, it is possible only to speculate as to the range of factors that differentially affect migration patterns across cities. One possibility is that restrictive changes in provincial social welfare policies introduced during the early 1990s in Alberta and Ontario affected migration patterns in major cities in those provinces. Similarly, the role that certain cities play as primary service centres for surrounding Aboriginal settlements (such as Thunder Bay and Winnipeg) is a critical feature affecting migration.

Speculation aside, an important point to note with respect to migration to and from CMAs is not so much the impact of net migration, which is relatively small for the 1991–1996 period, but rather the "churn" represented by the relatively high rates of in- and out-migration, especially for the Registered Indian population. Again, reserves play a distinct role in the difference between Registered Indians and other Aboriginal people in their migration patterns to and from CMAs. As a point of both origin and destination, reserves serve to increase the "churn" to and from cities. This can be demonstrated in the case of the Winnipeg CMA. Between 1991 and 1996, some 3,500 Registered Indians migrated to Winnipeg, yielding an in-migration rate of 200 migrants per 1,000 Registered Indians in Winnipeg, more than twice as high as the rates of other Aboriginal groups and other Canadians. Similarly the out-migration rate of Registered Indians was 210 per 1,000, again notably higher than the corresponding rates of 110 and 100 for "other Aboriginal" and Canadians respectively. A large part of this difference in rates between Registered Indians and other groups was due to movement to and from reserves. Twenty-seven per cent of Registered Indian migrants moving to Winnipeg during this period came from reserves, while some 47 per cent, of Registered Indians moving from Winnipeg were moving to reserves. Clearly, if the flows between Winnipeg and reserve communities were removed, the rates of Registered Indians migration into and out from Winnipeg would be much closer to those observed for other populations.

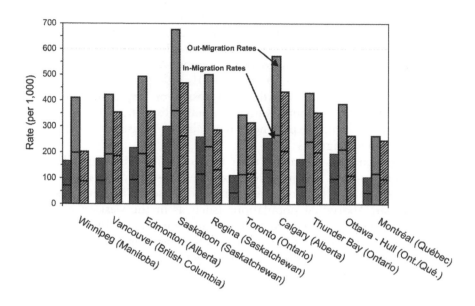

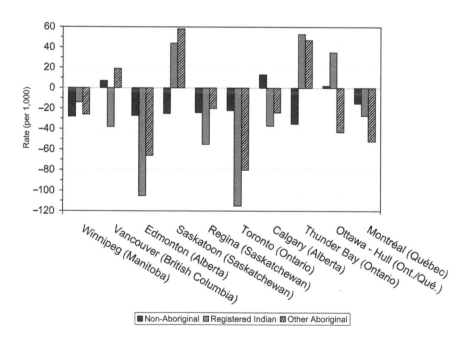

Figure 7.8 Five-year gross and net migration rates for top ten cities, 1991–1996.

Source: Norris *et al.* 2001: 43–4. Reproduced with the permission of the Minister of Indian and Northern Affairs, 2002.

Factors influencing migration

Several factors have been identified as contributing to individual and group decisions to migrate from reserve communities. Chief among the push factors has been the lack of employment opportunities, and resulting poverty and difficult social conditions in many communities (Falconer 1985; Hawthorn 1966; McCaskill 1970; Trovato *et al.* 1994). However, social factors such as marriage and the desire for a less restrictive lifestyle among young people have also been cited as contributory factors (Krotz 1990). More detailed analyses of social and economic conditions on reserves reveal that a complex interaction of community development issues compel migration. For example, Gerber (1984) found that the quality of housing in reserve communities was an important predictor of migration levels, as was institutional completeness, or the ability of a community to satisfy people's commercial, economic, and other needs without their having to leave the community. Some of these needs may include access to health facilities and educational opportunities on-reserve, with the lack of such facilities considered an important push factor (Trovato *et al.* 1994: 18). Other analysts have suggested that governance issues in communities, in particular perceptions of inequitable distribution of benefits and housing among community members, could also play an important role in migration decision-making (Cooke 1999).

The movement from cities to reserves has generally been described as return migration, or movement to reserves by people who had once left them (Frideres 1974; Siggner 1977; M.J. Norris 1990). This migration is often characterized as resulting from an inability of people who have left reserve communities to subsequently find employment or otherwise adjust to life in the city (Trovato *et al.* 1994: 287). Certainly, one important push from cities to reserves is a lack of access to affordable urban housing stock (Cooke 1999; Trovato *et al.* 1994: 28). While many reserve communities also face housing shortages, for those who are able to secure housing on-reserve, returning home may be preferable to remaining in the city, where affordable housing is often located in generally impoverished neighbourhoods.

Thus, rather than concentrating on factors that may push people out of cities to Aboriginal communities, it is also necessary to consider the pulls that may be presented by these communities. In this context, it is important to recognize that reserve communities represent a home base to which return is possible and relatively easy (Lurie 1967). In effect, reserves provide a "cultural hearth" and a critical mass of family and friends, cultural activities, and culturally appropriate services that may not be available off reserves. These factors make reserves attractive destinations for Aboriginal city dwellers, especially for those who are marginalized in urban environments. There is evidence that the support of extended families on reserves represents an important form of social capital, and one that is relatively unavailable in the city. Accordingly, reserve communities may be perceived as offering a better quality of life in terms of closer ties to family and other community members, lower crime rates, and better opportunities to participate in cultural activities. Some reserve communities have been described by residents as better places than urban centres in which to raise children given their generally lower crime rates and fewer problems of alcohol and drug abuse. At the other end of the life cycle, retirement to reserve communities has also been perceived as a desirable option (Cooke 1999).

Further indication of the reasons for migration is provided by the 1991 Aboriginal Peoples Survey (APS) (Clatworthy and Cooke 2001). According to this survey, the major reasons given for moving off a reserve (on to off) are issues regarding family (34 per cent), education (27 per cent), and housing (25 per cent), as shown in Figure 7.9. Surprisingly, only a small percentage cited employment reasons for such a move. For those Registered Indians moving back to a reserve (off to on) the major reasons were similar. The third group (off to off) provided reasons for their moves that were quite different from the other two groups in that employment (23 per cent) played a major role. For all three groups, family-related issues were the number one reason given as to why people moved. Also, nearly a quarter of the moves in all three groups were related to improved housing.

Some differentiation by gender is discernible in these reasons for movement. For example, Peters (1994) has suggested that Aboriginal women

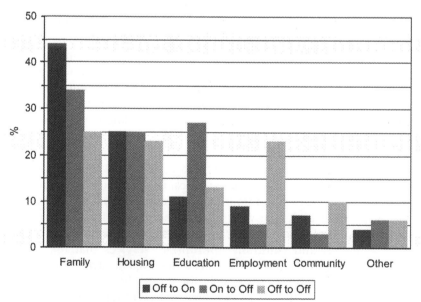

Figure 7.9 Reasons for migration from the 1991 Aboriginal Peoples Survey.
Source: Clatworthy and Cooke 2001.

tend to move in a family context, whereas men may tend to move as lone, "economically motivated" individuals. Further evidence in support of this is available from women migrants to prairie cities who are more likely to cite housing or family reasons, or problems with their home communities, as reasons for migration. Men, on the other hand, tend to report that they had moved for economic reasons (Clatworthy and Hull 1983). More recent evidence from the 1991 APS indicates that males were more likely than females to cite housing factors and education as reasons for leaving reserve communities, whereas females were much more likely to identify family-related issues (Clatworthy 2002). In the opposite direction, men tended to indicate housing and employment as reasons for moving from off to on reserve, in contrast to females who again were more likely to indicate family and community reasons. In each case, however, propensity to move is heavily influenced by stages in the life cycle and personal attributes, such as educational attainment and attachment to traditional culture as indicated by retention of an Aboriginal language.

While individual characteristics are important factors in stimulating migration between reserves and other areas, the social, geographic, and economic characteristics of reserve communities, including their population size and proximity to urban areas, are also important predictors of out-migration (Gerber 1984). Also noted is the fact that that communities nearest to urban service centres (within 60 km), and those furthest from urban centres (more than 300 km), have higher predicted out-migration rates and lower in-migration rates than do communities at more moderate distances (Clatworthy and Cooke 2001). It may be that people living in urban areas are able to maintain relationships with home reserve communities that are nearby, without undertaking return moves to those communities. In the case of distant communities, on the other hand, it would seem that the additional cost of moving may reduce return migration, while access to services provided in urban areas far away requires out-migration from the community.

Smaller reserve communities also tend to have higher predicted in- and out-migration rates (Clatworthy and Cooke 2001), which may reflect a relative lack of access to services and institutions (Gerber 1977, 1984). In addition, remote reserves with small populations tend to have the highest out-migration rates, suggesting that larger remote communities are better able to retain potential migrants. On the other hand, the level of economic development in a reserve community tends to increase, rather than decrease, mobility out of the community, suggesting that interconnectedness with the off-reserve labour market and the broader economy may be a stimulus to out-migration (Clatworthy 2002).

In contrast to this, reserve communities with higher rates of participation in traditional cultural activities tend to have lower rates of out-migration, perhaps indicating that social cohesion in reserve communities is an important influence on migration. In support of this idea, a positive relationship has been found between frequent turnovers of community leadership and out-migration, based on models using APS data (Clatworthy and Cooke 2001). Clearly, migration results from the interplay between personal characteristics of potential movers, the characteristics of communities of residence, and those of potential destinations.

Policy implications and responses

The migration and mobility patterns of Registered Indians described above have important policy implications for both urban and reserve communities (Ponting 1997). The most obvious is that, rather than the gradual loss of population that had been predicted in the 1960s, reserve communities have been gaining population due to migration. The significance of this from a policy perspective is that it adds to the already considerable need for additional employment, housing and related infrastructure that is produced by high rates of natural increase (Clatworthy *et al.* 1997; Loh *et al.* 1998). At the same time, empirical evidence exists to suggest that out-migration could increase in those communities where economic development results in stronger linkages to the broader economy and off-reserve labour market (Clatworthy 2002).

The most immediate consequences for policy, however, arise from the bi-directional movement between reserves and urban areas, combined with high rates of mobility among Registered Indians living off-reserve. The implication of this pattern of mobility is not so much change in the size of urban or reserve populations, but high population turnover in many communities, both rural and urban, with attendant disruptive effects on individuals, families, communities and service providers. In this regard, it should be noted that Registered Indians, who are subject to the provisions of the Indian Act, tend to have higher mobility than do Métis, Inuit, or non-status Indians, and reserves tend to have higher in- and out-migration than do other Aboriginal communities. While this is suggestive of the importance of on-reserve benefits in influencing mobility patterns, the characteristics of Aboriginal communities, urban areas, and those of individual migrants clearly also affect movement decisions. Reserve communities in Canada differ widely in their economic, socio-cultural, and geographic characteristics, and migration flows between individual communities and Canadian cities are the outcomes of particular sets of circumstances that can only be properly understood by examining them individually, as well as within the general Canadian context.

The implications of the patterns of mobility we have described include difficulties for the delivery of social services to Aboriginal populations. Thus, programmes that identify their clientele by neighbourhood, notably in the areas of health, employment and education, are particularly affected. As one example of this, Clatworthy (2000) has found a strong positive relationship between the Aboriginal share of population and school turnover rates in central Winnipeg neighbourhoods such that a 10 per cent increase in the Aboriginal share of neighbourhood population results in a 14 per cent increase in student turnover. While the authors are not aware of any specific studies documenting the consequences of such a relationship, there is considerable anecdotal evidence suggesting that student perfor-mance is adversely affected by high turnover. Negative impacts of high mobility on service delivery may also be heightened for certain sub-groups in the population. For example, female lone parents are among the most mobile, and yet are often the most in need of access to social services. Again, while anecdotal accounts of these effects exist, systematic research into the consequences of population turnover remains to be undertaken.

While mobility and migration may affect programme delivery, it is also the case that policies and programme delivery may affect migration. Most federal government programmes for Registered Indians apply only to those living in reserve communities, while Registered Indians living off-reserve generally access services through provincial and municipal govern-ments, as does the Canadian population in general. Despite the evidence of net inflows to reserves, Reeves and Frideres (1981), and Bostrom (1984) have claimed that through the 1970s, the federal government actively curtailed programmes available on reserves, in order to encourage migration off reserves, and to reduce its fiscal obligations. Reeves and Frideres (1981) further claim that the decreasing effectiveness of public service delivery to a growing urban Aboriginal population has led to less effective acculturation into urban communities, resulting in return migration to reserves. However, these concepts are difficult to operationalize, and there has been no real evidence yet presented that the implementation of a policy to promote out-migration from reserves, or that better service delivery in urban areas would reduce return migration.

One particular policy challenge related to the migration of Registered Indians is the provision of housing in both reserve and off-reserve areas. Housing is one of the primary reasons for mobility on- and off-reserve. On reserves, the funding of capital grants and operating subsidies to housing is a responsibility of the federal government. However off reserves, the matter is more complicated owing to the regulatory jurisdiction of the provincial and territorial governments and the changing nature of housing programmes over the years. The federal government has provided, and continues to provide, operating subsidies to existing social housing for low-income Canadians on a cost-shared basis with provincial and territorial governments. While the precise arrangements vary from province to province, the administration of this housing has been, or is in the process of being, entirely transferred to them. Funding for new social housing was curtailed in 1993 as a cost-cutting measure by the federal government and most provinces joined suit, although a number of renovation programmes continue.

In the 1970s and early 1980s, the federal government unilaterally funded some social housing directly with non-profit organizations; one stream of this was urban Aboriginal housing. Most of this housing is included in the transfer to the provinces. As such, there is no urban Aboriginal housing policy at the national level, rather policy and administration is decentralized, falling under the domain of provincial and, where further delegated, municipal governments. To some extent this reflects the diversity of regional and local housing needs off-reserve. As a consequence, policies and programmes concerning urban Aboriginal housing are not developed in a consistent fashion across cities. Thus, municipal governments, who often deal directly with issues such as housing in the delivery of *ad hoc* programmes and services, must do so in the absence of a policy framework (Hanselmann 2001; personal correspondence, P. Deacon, Research Division, Canada Mortgage and Housing Corporation).

One of the key components contained in the federal government's response to the recommendations of the 1996 Royal Commission on Aboriginal Peoples (Department of Indian Affairs and Northern Development 1997) is a renewed federal commitment to improving the conditions of all Aboriginal populations, including Registered Indians living on- and off-reserve, as well as non-status, Métis, and Inuit people. This federal policy has included commitments to improving housing and social services, as well as to building economic capacity in reserve communities. The federal government has also undertaken discussions with Métis and urban Aboriginal organizations with the intention of developing new models of funding, and policy-making, as well as approaches to self-government. Such institutional arrangements could affect population movement and might thereby reduce the level of population churn, whether or not this effect is intended.

While migration does not contribute to redistribution of the Aboriginal population as much as might be commonly thought, the fact that population mobility remains high is central to understanding many of the social, economic and political development issues that face Aboriginal people in Canada. Census data have consistently shown that the Registered Indian population off-reserve is highly mobile, moving to a greater extent than the general population, especially in the young adult age groups. Combined with the frequent mobility observed between reserves and urban areas, this has significant implications for the building of institutional completeness and capacity within all communities. Looking ahead, in the case of reserves, the shortage of functional housing and job opportunities in First Nation communities has the potential to increase pressures to migrate from reserves, especially against a background of projected rapid growth in the working-age population. On the other hand, these same projections point to an ageing Aboriginal population and this process may serve to reduce transience as people have less inclination to relocate at older ages. As for the present, it is the frequency of population movement between reserves and cities, not an exodus from the former, that has the greatest implications for the well-being of Aboriginal people and communities.

Notes

1 The views expressed in this chapter are those of the authors and do not necessarily represent the views of the Department of Indian Affairs and Northern Development (DIAND). The authors would like to acknowledge with thanks technical support provided by Gerry Ouellette of Statistics Canada and Lucette Dell'Oso of the Research and Analysis Directorate, DIAND; and special thanks to Doug Norris of Statistics Canada for facilitating access to data.

2 The Indian Act affects counts of both registered Indians and non-status Indians. In 1985, amendments were made to the Indian Act (Bill C-31) that restored Indian status (reinstatements) to those persons who had lost status as a result of provisions in earlier versions of the act (such as the out-marriage of registered Indian women to non-status men).

3 Métis traditionally refers to those persons of Aboriginal ancestry who are descended from the historic Red River Métis community of Western Canada. Now it also refers to persons who report Métis ancestry or identity, as distinct from Indian or Inuit.

4 Defining Aboriginal populations is a multi-dimensional phenomenon involving overlapping concepts of ethnic affiliation and legal status. Accordingly, the separate counts shown here include multiples. For example, some Métis have legal Indian status.

5 In 1985, amendments were made to the Indian Act (which are commonly referred to as Bill C-31). These amendments contained three sets of provisions that play a central role in shaping Indian demography: the reinstatement of Registered Indian status to individuals who had lost status through prior versions of the Act and for the first time registration of their children; new rules governing entitlement to Indian registration for all children born to a Registered Indian parent after 17 April, 1985 (i.e. status inheritance rules); and the opportunity for individual First Nations to establish their own rules and provisions governing membership (i.e. band membership rules).

6 It should be noted that our interpretation of migration is limited by the presence of the phenomenon of ethnic mobility. For example, if a person moves from a reserve to a city and decides not to declare himself as Aboriginal in the census, then that person would of course not be included in the count of Aboriginal out-migrants from reserves. This interaction between migration to cities and ethnic mobility to the non-aboriginal population has generated a growing number of "invisible urban Aboriginals". This situation would have occurred mostly in previous censuses. More recent censuses show a significant growth of Aboriginal populations in urban centres, which indicates that the "invisible urban Aboriginals" made themselves "visible" again by self-reporting as Aboriginal. See Guimond (forthcoming). Therefore it is important to stress here from this perspective that we are looking at the migration patterns of those individuals who are identifying as Aboriginal (including Registered Indian) at the time of the census.

7 The demographic impact of the reinstated population is understated here because children of Registered Indians, who were reinstated under Bill C-31, that were born since 1985 are not counted as reinstated Indians. In a recent study, Clatworthy (2001) estimates that between 1985 and 1999, about 59,800 children have been born who qualify for registration due to the new rules.

8 Note that figures between 1991 and 1996 are not directly comparable for a number of reasons, the major one being differentials in incomplete enumeration of reserves.

9 Not adjusted for incompletely enumerated reserves.

10 Includes the Métis, the non-status Indians and the Inuit.

References

Bostrom, H. (1984). 'Government policies and programs relating to people of Indian ancestry in Manitoba', in R. Breton and G. Grant (eds) *Dynamics of Government Programs for Urban Indians in Prairie Provinces*, Montreal: Institute for Research on Public Policy.

Chapman, M. (1978) 'On the cross-cultural study of circulation', *International Migration Review* 12(4): 559–69.

Clatworthy, S.J. (1996) *The Migration and Mobility Patterns of Canada's Aboriginal Population*, Report prepared for the Royal Commission on Aboriginal Peoples, Ottawa: Canada Mortgage and Housing Corporation, and the Royal Commission on Aboriginal Peoples.

Clatworthy, S.J. (2000) 'Patterns of residential mobility among Aboriginal peoples in Canada: an urban perspective', Presentation made at the Aboriginal Strategy Federal Workshop, Regina, Saskatchewan, May 11–12, 2000.

Clatworthy, S.J. (2001) *Re-Assessing the Population Impacts of Bill C-31*, Report prepared by Four Directions Project Consultants for the Research and Analysis Directorate, Department of Indian Affairs and Northern Development.

Clatworthy, S.J. (2002) *Registered Indian Migration between On- and Off-Reserve Locations, 1986–1996: Summary and Implications*, Report prepared by Four Directions Project Consultants for Research and Analysis Directorate, Department of Indian Affairs and Northern Development.

Clatworthy, S.J. and Cooke, M. (2001) *Reasons For Registered Indian Migration*, Report prepared by Four Directions Project Consultants for the Research and Analysis Directorate, Department of Indian Affairs and Northern Development.

Clatworthy, S.J. and Hull, J. (1983) *Native Economic Conditions in Regina and Saskatoon*, Winnipeg: Institute of Urban Studies.

Clatworthy, S.J., Hull, J. and Loughren, N. (1997) *Implications of First Nations Demography*, Report prepared by Four Directions Consulting Group for the Research and Analysis Directorate, Indian and Northern Affairs. Ottawa: Department of Indian Affairs and Northern Development.

Cooke, M.J. (1999) 'On leaving home: return and circular migration between First Nations and Prairie Cities', unpublished Master's thesis, University of Western Ontario.

Department of Indian Affairs and Northern Development (1997) *Gathering Strength: Canada's Aboriginal Action Plan*, Ottawa: Department of Indian Affairs and Northern Development.

Falconer, P. (1985) 'Urban Indian needs: federal policy responsibility and options in the context of the talks on Aboriginal self-government', unpublished discussion paper, Winnipeg: Institute of Urban Studies.

Frideres, J.S. (1974) 'Urban Indians', in J.S. Frideres (ed.) *Canada's Indians: Contemporary Conflicts*, Scarborough: Prentice-Hall.

Gerber, L.M. (1977) 'Community characteristics and out-migration from Indian communities: regional trends', paper presented at the Department of Indian Affairs and Northern Development.

Gerber, L.M. (1984) 'Community characteristics and out-migration from Canadian Indian reserves: path analyses', *Canadian Review of Sociology and Anthropology*, 21(2): 46–54.

Guimond, E. (1999) 'Ethnic mobility and the demographic growth of Canada's

Aboriginal population from 1986–1996', in *Report on the Demographic Situation in Canada, 1998–1999*, Ottawa: Statistics Canada.

Guimond, E. (forthcoming) 'L'explosion démographique des populations autochtones du Canada de 1986 à 1996', Université de Montréal, département de démographie, texte soumis en vue de l'obtention d'un diplôme de doctorat.

Hanselmann, C. (2001) *Urban Aboriginal People in Western Canada: Realities and Policies*, Report by the Canada West Foundation, Calgary: Canada West Foundation.

Hawthorn, H.B. (ed.) (1966) *A Survey of the Contemporary Indians of Canada*, Ottawa: Indian Affairs Branch.

Howes, C. (2000) 'The new Native tycoons', *National Post*, 27 January: D05.

Krotz, L. (1990) *Indian Country: Inside Another Canada*, Toronto: McClelland and Stewart.

Loh, S., Verma, R., Ng, E., Norris, M.J., George, M.V. and Perreault, J. (1998) 'Population Projections of Registered Indians, 1996–2021', unpublished report prepared by the Population Projections Section, Demography Division, Statistics Canada for the Department of Indian Affairs and Northern Development.

Lurie, N.O. (1967) 'The Indian moves to an urban setting', in *Resolving Conflicts – A Cross-Cultural Approach*, Winnipeg: University of Manitoba Extension and Adult Education Department.

McCaskill, D.N (1970) 'Migration, adjustment, and integration of the Indian into the urban environment', unpublished Master's Thesis, Carleton University.

Melting Tallow, P. (2001) 'A place to hang your headdress', *Aboriginal Times*, 10(5): 31–4.

Norris, D.A. and Pryor, E.T. (1984) 'Demographic change in Canada's north', in *Proceedings – International Workshop on Population Issues in Arctic Societies*, Gilbjerghoved, Gilleleje, Denmark, 2–5 May, 1984.

Norris, M.J. (1985) 'Migration patterns of status Indians in Canada, 1976–1981', paper prepared for the Demography of Northern and Native Peoples in Canada session at Statistics Canada, June, 1985.

Norris, M.J. (1990) 'The demography of Aboriginal people in Canada', in S.S. Halli, F. Trovato, and L. Driedger (eds) *Ethnic Demography: Canadian Immigrant, Racial and Cultural Variations*, Ottawa: Carleton University Press.

Norris, M.J. (1992) 'New developments and increased analytical possibilities with mobility and migration data from the 1991 Census', paper prepared for the annual meeting of the Canadian Population, Charlottetown, Prince Edward Island, 2–5 June, 1992.

Norris, M.J. (1996) 'Contemporary demography of Aboriginal peoples in Canada', in D.A. Long and O.P. Dickason (eds) *Visions of the Heart: Canadian Aboriginal Issues*, Toronto: Harcourt Brace Canada.

Norris, M.J. (2000) 'Aboriginal peoples in Canada: demographic and linguistic perspectives', in D.A. Long and O.P. Dickason (eds) *Visions of The Heart: Canadian Aboriginal Issues* (2nd edition), Toronto: Harcourt Brace Canada.

Norris, M.J., Beavon, D., Guimond, E. and Cooke, M. (2001) *Registered Indian Mobility and Migration: An Analysis of 1966 Census Data*, Ottawa: Indian and Northern Affairs Canada.

Peters, E. (1994) *Demographics of Aboriginal People in Urban Areas, in Relation to Self-Government*, Ottawa: Department of Indian Affairs and Northern Development.

Ponting, J.R. (ed.) (1997) *First Nations in Canada: Perspectives on Opportunity, Empowerment, and Self-Determination*, Toronto: McGraw-Hill Ryerson.

Reeves, W. and Frideres, J. (1981) 'Government policy and Indian urbanization: the Alberta case', *Canadian Public Policy*, 7(4): 584–95.

Robitaille, N. and Choinière, R. (1985) *An Overview of Demographic and Socio-Economic Conditions of the Inuit in Canada*, Ottawa: Indian and Northern Affairs Canada.

Siggner, A.J. (1977) *Preliminary Results from a Study of 1966–71 Migration Patterns among Status Indians in Canada*, Ottawa: Department of Indian Affairs and Northern Development.

Statistics Canada (1998) *Mobility and Migration: 1991 Census*, cat no. 93–322, Ottawa: Statistics Canada.

Trovato, F., Romaniuc, A. and Addai, I. (1994) *On-and Off-Reserve Migration of Aboriginal Peoples in Canada: A Review of the Literature*, Ottawa: Department of Indian Affairs and Northern Development.

Young, T.K., Reading, J., Elias, B. and O'Neil, J.D. (2000) 'Type 2 diabetes mellitus in Canada's First Nations: status of an epidemic in progress', *Canadian Medical Association Journal*, 163(5): 561–6.

Part III
Local contingency

8   The politics of Māori mobility

Manuhuia Barcham

Māori society was transformed in the second half of the twentieth century by an almost wholesale movement from rural to urban areas. In 1936 more than 80 per cent of Māori lived in rural areas – 50 years later more than 80 per cent of Māori were living in urban centres. This demographic shift prompted dramatic changes in the structure of many Māori social institutions, and much recent work has been concerned with exploring the implications of these changes. However, few analyses have explicitly sought to explore the linkages that exist between Māori migration and Māori political, social and organizational change, and none have specifically considered the role played by government policy. In line with approaches to the study of population mobility that stress the importance of understanding institutional and organizational influences (Swindell and Ford 1975), this chapter sketches out the nature of such linkages and then contextualizes its findings using the experiences of one tribe from the North Island of New Zealand – Ngāti Kahungunu – focusing particularly on the recent reconstruction of tribal structures and the related phenomenon of return-migration.[1]

Early Māori mobility patterns

Traditional Māori society is typically viewed as being organized along a framework of kin-based descent groups with three major social classificatory units, *Whānau* (immediate and extended family), *Hapū* (clan), and *Iwi* (confederation of Hapū). During the early pre-contact period the Māori population was highly mobile, with extensive trade networks woven between the territories of these various groupings. The subsequent "classical" period of Māori society saw population mobility increase as endemic warfare, due in large part to a shrinking resource base especially in terms of food (particularly meat), led these groups to engage in increased levels of warfare and conquest. This warfare-induced mobility intensified with the arrival of the first Europeans as the introduction of the musket added a new degree of lethality to these conflicts. The period of the early nineteenth century was thus characterized by extremely high levels of mobility for

Māori as large numbers of people were displaced as they attempted to escape the various conflicts that raged over the country during this period. The strong genealogical links that some of the Iwi of the East Coast of the North Island have with various Iwi from the inner North Island are a direct result of these frenzied population movements. However, with the extension of the colonization process following the signing of the Treaty of Waitangi in 1840 these significant population movements began to abate as European rule of law became increasingly dominant.

While some degree of movement within the Māori population continued it was mainly localized, involving limited forms of circular migration – visiting kin, trading in towns and pursuing traditional food gathering techniques – and there was little, if any, permanent movement of Māori to the country's urban settlements.[2] The expansion of European colonization in the wake of the signing of the Treaty of Waitangi also led to a decline in the political importance of traditional forms of Māori socio-political organization. Increased levels of European migration to New Zealand and the dramatic loss of the military power of Iwi and Hapū in the 1860s meant that by the end of the nineteenth century these Māori organizational structures found themselves in possession of an ever decreasing power base (Maaka 1997: 3).[3] Despite these dramatic changes, the Māori people continued throughout the nineteenth century to live mainly in the country's rural areas.

The beginning of the twentieth century saw little change to Māori residential and mobility patterns. The dominant European, and hence government, view of the day was that urbanization was something to be discouraged for Māori – as they, as a race, were perceived as being too backward to cope with the civilities necessary for city life. Despite this, at the beginning of the twentieth century small numbers of Māori began to migrate to the cities looking for work. Still convinced of the far-reaching negative consequences that this nascent migratory trend would have on Māori, the New Zealand government of the 1920s began to actively pursue policies to ensure that Māori remained in the rural environment. The establishment of the Native Trustee in 1920, was one of the first of these new policies, with one of its main purposes being to facilitate and promote Māori agriculture through the establishment of lending policies for Māori farmers. This policy trajectory was consolidated further when, in 1927, the merging of Māori land interests became official government policy. It was thought that this consolidation of land titles would help ensure the success of the long-term government policy of maintaining Māori in rural areas by facilitating the creation of blocks large enough to be farmed economically. These efforts reached their logical conclusion in 1929 when the government passed legislation authorizing the state to develop Māori land (Butterworth 1991a: 28). Thus, the ultimate outcome of policies aimed at keeping Māori out of urban environments was the creation of state-funded Māori farming.

Notwithstanding such policies, the outbreak of war in 1939 meant that Māori labour was increasingly required in the country's cities. While many Māori remained in the rural environment, as agriculture was a vital part of the war effort, many did move to the city in response to the government's new manpower regulations in order to work in vital industries such as metal-working. Thus during the 1936–1945 intercensal period many Māori, both men and women, moved to the cities as a result of the government's wartime policies of manpower relocation. Interestingly, the end of World War II did not see the abatement or even diminution of these Māori rural–urban migration flows. Instead the period immediately following the conclusion of the war witnessed a dramatic increase in the rate of Māori rural–urban migration flows. An important government policy development for Māori at this time was the passage in 1945 of the Māori Social and Economic Advancement Act. This Act established a network of regional Māori committees to provide input into Māori community development. The interesting point about these committees was that they were based on regional populations, not tribal affiliation – indeed they often cut across tribal territories. This Act is thus significant in that it was one of the first clear legislative indications of the preference of government to deal with Māori as a race, as opposed to as a tribal people.

Rural–urban migration

The post-World War II exodus of Māori to urban centres can be explained by a number of interrelated factors. In contrast to the employment lows of the Great Depression, the period following World War II was characterized by dramatic labour shortages – with the manufacturing and service industries in particular growing at a faster rate than the non-Māori urban labour market could supply. Additionally, the enormous growth in the Māori population between 1921 and 1951 meant that by the 1940s the majority of Māori communities were beginning to suffer acute over-population in relation to their land resources (Butterworth 1991a: 34). Also, the subdivision of Māori land had proceeded to the extent that many farms were barely economic and were unable to provide the standard of living that Māori families desired at the end of World War II (Butterworth 1991a: 35). Given the lure of better employment and income opportunities in the city, it is not surprising that many Māori left the land. These factors – combined with the generally slow rate of regional development in those parts of New Zealand most heavily populated by Māori – all acted to ensure that by the middle of the twentieth century Māori were moving to urban areas at an unprecedented rate.[4]

Overall, the New Zealand government was slow to adjust to the fact of Māori rural–urban migration. Accordingly, the Department of Māori Affairs continued through the 1950s and early 1960s to direct its main efforts towards the establishment of Māori farms.[5] However, despite the inattentiveness of government to the changing situation of Māori, a number of forward-thinking ministers of Māori Affairs did manage to initiate policies that were more attuned to the realities of growing Māori urbanization. One such directive involved the building of new hostels for Māori immigrants in cities and the refurbishment of existing hostels (Butterworth 1991a: 36). Hostels for Māori had existed in various urban areas around the country since the nineteenth century. Initially established for Māori travellers passing through urban areas, the new plan was that with increasing numbers of young Māori moving to the cities looking for work these hostels offered an opportunity to socialize Māori into urban norms and to provide young migrants with a place to stay while they adjusted to their new environment. In retrospect, these hostels provided the environment in which many young Māori first mixed with other young Māori from different tribal backgrounds – a process important in the later creation of a pan-Māori identity.

Despite an unfolding demographic transition – for this is where the migration flows were leading – the political shift to a conservative government in the 1950s resulted in renewed focus on policies of land development and title reform for Māori land in a final attempt to maintain Māori in a rural environment. This situation reversed in 1957, however, with the return of the New Zealand Labour Party to office. In response to the massive changes occurring within Māori society as a result of this substantial rural–urban migration during this period, the Labour government developed specific programmes of relocation and urban housing development for Māori – programmes given theoretical credence by the publication of the Hunn Report.[6] Released in 1960, this report signalled a dramatic shift in the way that the New Zealand government viewed the process of Māori urbanization. In contrast to earlier policies, which had perceived the issue of Māori urbanization as a problem to be overcome, this report signalled a shift towards a perception that the movement of Māori to cities was a highly desirable phenomenon which needed to be actively promoted as a way to bridge the gaps in health, employment and education that were preventing the "integration" of Māori into wider New Zealand society (Hunn 1960: 14).

The 1960s thus saw the New Zealand government become committed to a policy of active relocation of Māori from areas where opportunities for employment were limited (which meant, in the discourse of the day, rural areas) towards areas where work was bountiful (which in this case meant the country's urban areas). The welfare division of the Department of Māori Affairs thus went out of its way to actively stress the advantages of life in the city to young Māori and their families through promotional campaigns, while increasing levels of finance were made available through the Department of Māori Affairs for Māori migrants to purchase houses in the cities. Continued high migration levels meant that the Department of Māori Affairs was building up to 1,000 houses a year and obtaining loan finance and rental accommodation from the State Advances Corporation to make available a further 500–700 houses for Māori migrants (Butterworth 1991a: 38).

The result of this policy shift was that by the end of the 1960s over 1,200 Māori families and thousands of individuals had been moved to urban centres, with 92 per cent of the net internal migration of the Māori population during the 1961–1966 period being to urban areas (Poulsen and Johnston 1973: 151–72). Net Māori migration in the 1950s and 1960s was thus dominated by outflows from the rural areas of the northern, central and eastern areas of the North Island with corresponding inflows to the main urban centres. This movement had two distinct components. The first, and main, component comprised whole households from poor housing, low-income rural areas with low opportunities for jobs. The second subsidiary flow component comprised young, single people moving long distances for higher-income jobs (Poulsen *et al.* 1975: 315). It was during this era then that Māori made the demographic switch from being a predominantly rural people into a predominantly urban people. In 1961, the majority of Māori (54 per cent) still lived in rural areas but by 1966 the majority (66 per cent) of Māori were living in an urban environment (Pool 1991: 153).

As would be expected, one of the significant impacts of the extensive urbanization of the Māori population in the thirty years following the conclusion of World War II was a radical transformation of the Māori labour force. In 1956, the majority of Māori (26 per cent) were employed in the agricultural, forestry and fishing sectors, but by 1966 this percentage had shrunk to only 17 per cent. Similarly, in 1956 only 24 per cent of the Māori population worked in the manufacturing sector, yet only ten years later in 1966 a massive 32 per cent of the Māori population were working in industries within this sector, and by 1970 the manufacturing sector had become the largest net employer of Māori (Bu 1993: 191). Additionally throughout the 1960s the government continued its preference for dealing with Māori as a race and not as a tribal people. Thus, in 1962 the government passed the Māori Community Development Act which replaced the earlier Māori Social and Economic Advancement Act. This new Act extended the remit of the earlier Act by grouping local community committees into districts and then combining these districts so as to allow for the creation at the national level of a New Zealand Māori Council – further alienating the traditional tribal organizational structures of Māori society.

By the early 1970s, the rural exodus of Māori was beginning to diminish for several reasons. For one thing, departmental funding for the Ministry of Māori Affairs had been reduced from NZ$5.4m to NZ$1.2m, and so the Department no longer had the resources to maintain relocation policies as vigorously as had been the case in the 1960s. Additionally, Māori communities were no longer as convinced of the utility of sending all their young people to the cities.[7] These factors combined with a slowing of the

New Zealand economy meant that over the period 1976–1986 no significant regional redistribution occurred within the Māori population, with
the flows and counter-flows into and out of the cities being relatively evenly
balanced (Parr 1988). In addition to this slowing down of Māori rural–
urban migration, the 1970s also witnessed the emergence of a new Māori
political dynamic. The economic and social advancement which the move
to the urban environment was supposed to deliver had generally not
materialized, and many Māori consequently felt that they had merely
swapped their position as unskilled rural workers for that of unskilled
urban workers, though with the added problem of a general loss of cultural
identity. The 1970s thus witnessed a growing sense of dissatisfaction among
many Māori with the place that they held in New Zealand society – a sense
of dissatisfaction which was felt most keenly within the younger Māori
population on the social margins of urban New Zealand.

Rising urban politicization

With hindsight, it is clear that the urbanization process had a profound
effect upon contemporary Māori society. Much of the population shift from
the 1940s to the 1960s involved young people in their late teens to early
thirties (Poulsen *et al*. 1975: 155), and their movement away from traditional tribal areas led to a breaking of traditional kinship ties and relationships. This severance from traditional social support mechanisms led urban
Māori to form their own. Thus, across the urban centres of New Zealand
young Māori migrants formed culture groups, met in sports clubs and
hostels, and congregated at church gatherings (Walker 1995: 501). Removed
from their tribal homes these young migrants mixed and interacted with
others from very different tribal backgrounds than their own, many sharing
only the common experience of being Māori in the city. An important
aspect of this socialization was the "imagining" of a wider pan-Māori
community. Urbanization thus led Māori, from many different tribal
backgrounds, to begin to share a common sense of place and identity. This
growth in solidarity was an important factor as it laid the base for the later
unity and coherence of their, and their children's, struggle against the
perceived oppressive domination of the New Zealand state. While the
increase in the absolute numbers of Māori in the cities was perhaps the
most important factor impacting upon Māori society in the post-war years,
there was also another important interrelated factor – the increase in the
number of Māori tertiary graduates. These new graduates, raised in the
urban environment and unsure of their position in New Zealand society,
were dissatisfied with the status of Māori and began in the late 1960s to
actively agitate for change. Indeed, the passage of the Treaty of Waitangi
Act 1975, which established the Waitangi Tribunal – a quasi-juridical
institution created to provide a conduit for Māori to pursue grievances
concerning the Treaty – was a direct result of this new protest movement.[8]

Māori politicization in the 1970s therefore emerged from a combination of the dynamics of post-World War II rural–urban migration and the related phenomena of the emergence of a new Western-educated Māori intelligentsia. This increased politicization meant that by the 1980s the New Zealand government was forced to re-evaluate the way in which the Māori population was approached in terms of policy construction and implementation.[9] Moving away from the use of the abstract category "Māori" the government began to increasingly identify Māori in terms of tribal affiliation. As such, one of the key developments in government policy in this period was its focus on the re-development of tribal structures in Māori society. This development, however, can also be seen as an extension of the recognition of the initial claims made by Māori leaders at the Hui Taumata – a meeting which owed much to the urban protest movements of the 1970s for its existence.

Convened jointly by the Ministers of Māori Affairs and Finance the Hui Taumata (Māori Economic Summit Conference) was held in 1984 shortly after the fourth Labour government took office and it was here that Māori leaders argued that as they understood Māori needs best, the best outcomes for Māori could be delivered if increased control of the resources allocated to Māori were placed under Māori control. Iwi development therefore became the preferred government policy focus for Māori development and led to the subsequent devolution of a number of government programmes to various tribal authorities. Mātua Whāngai, Maccess and Mana Enterprises – all derived from the original Tū Tangata remit – were all thus premised on the view that Māori were better situated to provide services to Māori (Durie 1998: 8).[10] In a third sense though, this process of devolution to tribes – for this is what the process eventually became – can be seen as an extension of neo-liberalism, the ideology that underpinned the actions of the fourth Labour government's term in office.[11] Government concern with increased Māori (meaning tribal) control of the resources allocated them could on a more cynical reading thus be seen as just an extension of this new, more general policy line of devolving responsibility away from the state and onto communities (Rose 1996).

One of the side effects of this new focus was that the resulting policies tended to favour rural communities. Thus under the Mana Enterprises scheme the small Ngāti Awa Tribe received the same level of funding as the Iwi organization providing for the entire South Auckland region (Butterworth 1991a: 43). Government reports released during this period outlined the government's new policy position as being that *rangatiratanga* (Indigenous rights) was to be exercised through Iwi, and that increased focus was therefore to be placed upon Iwi as the major player in Māori economic development. During this period then, New Zealand government policy played a pivotal role in, what has been referred to elsewhere as the "Iwi-isation of Māori society" (Barcham 1998: 305–6) – a process given legislative force through the codification in law throughout the 1980s of a number of specific legislative acts and the release of a number of important government reports. One of the most important pieces of government legislation in this Iwi-ization of Māori society was the Treaty of Waitangi Amendment Act (1985).

While this Act empowered Māori by allowing claims under the Treaty of Waitangi to be backdated to 1840, it also meant that the tribal groups, territories and institutional forms acknowledged by the law were those that existed in 1840. This legislative Act thus played an important part in defining modern Iwi as the legitimate descendants of Māori society in opposition to the perceived "inauthenticity" of modern associational forms of Māori institutional organization. In 1988, this process was extended further with the release first of the discussion document *Tirohanga Rangapū* (Ministry of Māori Affairs 1988a), followed soon after by the policy document *Te Urupare Rangapū* (Ministry of Māori Affairs 1988b). However, probably the single most important piece of legislation in this process was the Runanga Iwi Act (1990). The Rūnanga Iwi Act provided various criteria which Iwi bodies needed to fulfil in line with government accountability standards in order to act as a vehicle for the direct distribution of government funds to Māori organizations. This plan never came to fruition though as the incoming National Party government repealed the Act almost as soon as they came to power in 1990 and reconstituted the Ministry of Māori Affairs and the Iwi Transition Agency into one body, Te Puni Kōkiri (Ministry of Māori Development), a body without the former department's service delivery capacity. Yet despite the fact that the defeat of the fourth Labour government by the National Party saw the repeal of the Runanga Iwi Act less than a year after its creation, its legacy remained, a legacy of Iwi with strong centralized governmental structures.

The passing of these various acts to empower Māori through the recognition of their status as Indigenous peoples therefore meant that by the beginning of the 1990s the New Zealand government had reached the conclusion that only traditional kin-based Iwi were their Treaty partners (Barcham 2000: 141). While the Treaty of Waitangi Act 1975 and its amendment in 1985 acted to solidify Iwi claims of historical continuity and continued territorial integrity, the Runanga Iwi Act 1990 acted to provide Iwi with a strong political and administrative framework and identity. These two aspects meant that from a position of relative weakness and powerlessness throughout the majority of the twentieth century, by the end of the 1980s Iwi had become active political bodies, with an existence strongly rooted in the law of the land. The restructuring of the New Zealand economy that paralleled these legislative and policy changes also resulted in major changes to the demographic characteristics of the Māori work force. The massive programme of deregulation and reform that the New Zealand economy underwent in the 1980s resulted in widespread unemployment, especially in the low-skilled section of the labour market.

As this was the sector of the labour market that the majority of rural–urban Māori immigrants had come to occupy, this period of restructuring and reform had an especially dramatic effect upon the Māori population.

Throughout the 1970s and early 1980s the labour force participation rates of Māori men and women had regularly been higher than those of non-Māori, with the labour force participation rate of Māori men sitting somewhere between 80 per cent and 85 per cent for the thirty years from 1951 to 1981. However, following the restructuring of the New Zealand economy in the wake of Labour government neo-liberal reforms, the participation rate had fallen by the end of 1991 to 67 per cent – a drop of almost 20 per cent. A similar, although not so dramatic, trend was evident in the labour force participation of Māori women over the same period. Whilst the restructuring of the New Zealand economy over the 1980s impacted upon the entire population, Māori and Pacific Islanders were particularly hard hit, concentrated as they were in the primary and secondary sectors of the economy. By the early 1990s substantial job losses, particularly in the manufacturing sector, meant that large numbers of Māori were now unemployed. This wave of unemployment was to have massive ramifications on Māori demographic patterns, helping to initiate, when combined with the Iwi-ization policies of the government, a new Māori mobility dynamic – that of return-migration.

Return-migration

By 1986 an equilibrium between Māori urban and rural migration appeared to have been established. No significant redistribution of the Māori population appeared to be occurring, and the outflows and inflows from the urban centres appeared to be evenly balanced. This equilibrium, however, disguised two important facts. The first was that while the flows were evenly balanced, Māori nonetheless continued to exhibit higher levels of migration than the general population. The second, and more interesting fact, was the emergence of a previously unknown mobility dynamic within the Māori population. Despite the radical political and economic change throughout the 1980s and 1990s young Māori continued to move as part of the transition phase from school to work, away from rural and small urban centres and towards the cities of Auckland, Wellington, Canterbury and Otago – the main New Zealand centres for tertiary education, training and jobs (Te Puni Kōkiri 1998: 7–8).

At the same time, though, other sections of the Māori population were responding to the massive social and economic transformations of the 1980s and 1990s with the instigation of new patterns of mobility. While Māori migration had previously been dominated by the process of rural–urban drift with some movement accounted for by movement between urban centres, by the 1980s increasing numbers of Māori had begun to migrate away from urban centres – with the total numbers of these migrants almost equalling those who were leaving rural areas (Butterworth 1991b: 4). While this urban–rural migration was not peculiar to Māori – as similar counter-urbanization trends were apparent across the populations of the Western world[12] – what was unique to Māori were the destinations that these urban–rural migrants chose. A study based on data from the 1991 census suggested that these people were leaving urban centres in order to return to their own tribal areas (Boddington *et al.* 1993).[13] Rather than migrating to areas of high amenity value, or to areas with strong employment opportunities, these migrants were thus choosing to return to those areas with which they held traditional links.

This pattern of Māori return migration has been observed over the period of the last four censuses (Bedford and Goodwin 1997: 19; Boddington and Khawaja 1993; Butterworth 1991b: 4; Parr 1988; Pool 1991; Scott and Kearns 2000). First noticed in the 1981 census this migration trend has become increasingly pronounced over the last twenty years. Thus, by 1986 there was a small but significant movement of Māori to rural areas, with the numbers moving to rural areas almost equalling those leaving (Parr 1988: 129). This movement has been most pronounced in the northern regions of the country with Auckland experiencing the greatest outflow of Māori and Northland experiencing the greatest inflow over the 1986–1991 intercensal period (Statistics New Zealand 1994: 13). Although Auckland experienced a small net gain in the subsequent five-year period this was minimal compared to the net gains made by the less urbanized regions of Northland, Waikato, the Bay of Plenty and Nelson/ Marlborough (Table 8.1). Between 1991 and 1996, rural areas achieved an overall net gain of almost 6,400 Māori from urban centres.

Emergence of this new dynamic in the redistribution of Maori population begs questions regarding its cause. Is it a response to the changing dynamics of the New Zealand labour market? Is it due to the increased focus on Māori development through Iwi prominent throughout the last twenty years of New Zealand government policy? Or is it instead due to the effect of a feeling of cultural alienation amongst the country's urban Māori coupled with a renewed interest in "traditional" Māori culture? In analysing the available data the answer to this question would appear to be derived from a mixture of all these.

While some migrants are moving back to their traditional tribal areas in an attempt to escape the current downturn of the New Zealand labour market, this emergent mobility pattern contradicts theories which posit that urban–rural migration in general is an attempt by individuals and families to remain in the labour market (Mabbett 1988: 666) as many Māori undergoing this form of migration are actually moving to areas of higher unemployment.[14] Additionally, the reconstruction of traditional Māori social and political organizations over the last twenty years has meant that some Māori have "returned home" to help in this revitalization process and to ensure that their children and grandchildren are able to partake of the

Table 8.1 Regional net migration of the Māori population, 1986–1996

Region	Net migration	
	1986–1991	*1991–1996*
North Island		
Northland	1,668	471
Auckland	−1,620	321
Waikato	534	801
Bay of Plenty	1,524	1,026
Gisborne	−495	−417
Hawke's Bay	−549	−591
Taranaki	−228	−414
Manawatu/Wanganui	−30	−408
Wellington	−963	−1,065
South Island		
Nelson/Marlborough	240	591
West Coast	−90	−81
Canterbury	261	558
Otago	93	252
Southland	−345	−915

Source: Te Puni Kōkiri (1998: 7).

cultural, social and economic opportunities that these developments present. Thus while Māori return-migration is at least partially dependent upon labour market conditions it is also dependent upon certain cultural factors, with the most important of these being the relationship between Māori and their *papa kāinga* (traditional tribal territories).

Accordingly, this new mobility dynamic thus appears dependent on a combination of both push factors – downturn in manufacturing industry, restructuring of the public sector and increasingly high costs associated with urban living – and pull factors – of which the most important appear to be those associated with a desire to be a part of the renaissance of Māori tribal identity. These ideas are now explored through an analysis of the experiences of one specific tribe – Ngāti Kahungunu – focusing particularly on the processes of tribal regrowth and the new mobility trend of return-migration.

Case study: Ngāti Kahungunu

The traditional territory of Ngāti Kahungunu is located on the eastern section of the North Island of New Zealand (Figure 8.1). Stretching from the town of Wairoa in the north to the Wairarapa in the south, Ngāti Kahungunu possesses the second largest territory of any modern Iwi and the third largest Iwi population.[15] The fertile and easily cultivatable nature of the majority of Ngāti Kahungunu's territory meant that by the turn of the century the majority of the tribe's land had passed into non-Māori hands. However, the region's natural abundance also meant that large sections of Ngāti Kahungunu's population were able to gain employment in the primary production sector, particularly in the farming and agricultural sectors. During World War II many young Ngāti Kahungunu men journeyed overseas to fight in the Māori Battalion, and many others, including women, moved to the cities to help in the war effort. At the conclusion of hostilities some of these early migrants remained in the

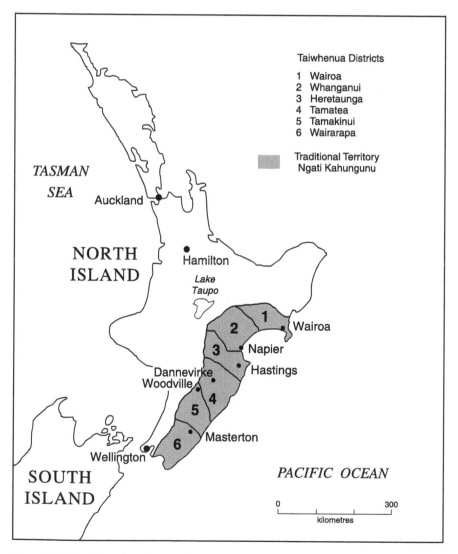

Figure 8.1 Traditional territory of Ngāti Kahungunu.

urban environment, although over the next twenty years this trickle of migrants was to become a flood.

Thus, in the 1950s and 1960s increasing numbers of young Ngāti Kahungunu began migrating to the country's large urban centres. While many left under the auspices of specific trade training apprenticeships organized by the Department of Māori Affairs, such as the carpentry apprenticeship scheme in Christchurch, many also left because of the lure of work and the prospect of an exciting time in the cities.[16] These initial migrants and their children played an important role in the establishment of the Māori protest movements of the 1970s with many individuals of Ngāti Kahungunu descent playing pivotal roles in their development. Throughout this period, however, Ngāti Kahungunu as a tribal structure was primarily a cultural institution invoked mainly on marae in formal speech-making occasions or in order to establish kinship links. Ngāti Kahungunu as a political body, as such, did not exist.

In the late 1970s, partly as a response to the social and political changes that the burgeoning Māori protest movement had brought about, a number of young Ngāti Kahungunu migrants and a number of Ngāti Kahungunu elders began to discuss the idea of a combined Ngāti Kahungunu Rūnanga (Council of Ngāti Kahungunu). This idea was then mooted at various meetings held throughout the tribal territory. In 1980, in response to this discussion, Ngāti Kahungunu elders at a meeting held at Omahu Marae then made a conscious decision to set up a tribal authority to unite and co-ordinate the activities of the people of Ngāti Kahungunu.[17] In 1981–1982 this idea was expanded to include the creation of a Rūnanganui O Takitimu to act as a parent body to administer and develop the affairs of all Māori living in the Ngāti Kahungunu tribal area and to co-ordinate the activities of the newly established Takitimu Rūnanga mo Te Reo (Takitimu Language Board). This Language Board had been set up in the late 1970s and early 1980s by the late Canon Wi Huata and his brother Ozzie in order to help promote the maintenance of the Māori language amongst the tribe's youth.[18] A large part of the impetus behind this move was the return of one of the Canon's sons to Hastings from Australia where he had moved in his early twenties and the discussions he had with his father concerning the need for positive steps to be undertaken in order to keep the traditions of Ngāti Kahungunu alive. This Language Board was divided into five Hau Wānanga (Districts Schools) which roughly followed the established boundaries of the local Māori councils.[19] Following the meetings of 1981–1982 the Te Rūnanganui o Takitimu (the Great Council of Takitimu) was established in 1983 and was based upon the same geographical boundaries as the District Schools of the Language Board: Wairarapa, Central Hawkes Bay, Hastings, Napier and Wairoa.

Fate or luck meant that this new Te Rūnanganui o Takitimu emerged as the New Zealand government began its movement towards neo-liberalism. As discussed earlier one of the key themes of neo-liberalism, as a technique of government, is the use of institutions such as the "community" to govern at a distance (Rose and Miller 1992). The devolution to Māori of greater control of their own governance, which the formation of the nascent Te Rūnanganui o Takitimu presaged, was thus in accord with the government's new ideological turn.[20] As a result the Rūnanganui was given a high degree of governmental legitimation, with Maccess and Mana schemes falling under their control as well as other more agency driven schemes such as Mātua Whāngai. The management of these schemes meant that the Rūnanganui was provided with government funding which enabled the nascent tribal organization to begin to develop a more sophisticated bureaucratic and administrative structure. These moves were further formalized in 1988 when the Te Rūnanganui o Takitimu was replaced by the incorporation of Te Rūnanganui o Ngāti Kahungunu (The Great Council of Kahungunu) – a move which firmly established the embryonic tribal institution as an Iwi authority under the terms of the government policy document *Te Urupare Rangapū*. With the signing of a pre-settlement Treaty Fisheries deal with the government in the early 1990s Ngāti Kahungunu were provided with a substantial monetary base. Unfortunately, a combination of inattention and mismanagement meant that by 1994 the monies from the initial settlement payment had been lost and the Rūnanganui was placed in receivership. The Rūnangunui was restructured and renamed Ngāti Kahungunu Iwi Incorporated. Throughout this process the basic tribal structure has not changed. Kahungunu's territory is split into five discrete *Taiwhenua* (regions) – which cover the same approximate area as the original Hau Wananga of the Takitimu Rūnanga mo Te Reo – with representatives from local marae being elected to represent their marae at the Taiwhenua level, and representatives from the Taiwhenua being elected to the Rūnanganui (Figure 8.1).

The continued growth and development of these new tribal structures has meant that additional structures of accountability have had to be created. Recent years have thus seen the development of a tribal register – and the creation of a permanent position within Ngāti Kahungunu Iwi Incorporated to deal with this issue. The consolidation of this modern tribal structure has also led to the establishment of regular elections involving members of Ngāti Kahungunu from as far abroad as Australia and England voting to elect their tribal representatives. This process has not been without its problems, however, and the most recent 2001 elections had a number of candidates disqualified for irregularities in the nomin-ation and voting process. These changes have come about due to a mixture of both internal and external pressures. Just as the accountability require-ments of the early Mana and Maccess schemes meant that Ngāti Kahungunu had to adopt a certain amount of bureaucratic sophistication, so too the ability to receive monies from various government compensation packages, most notably the fisheries settlement, requires that modern tribal structures have regular elections to ensure the continued existence of a strong mandate from the tribe's members. The development of Ngāti Kahungunu's "modern" socio-political structures have thus been directly influenced from their inception by the changes over the last twenty years in the governments policy trajectory with regards Māori. These changes have also, however, had a considerable impact upon Māori mobility patterns, and it is towards these new patterns that my attention now turns.

Returning to the theme of migration, what role have return-migrants played in the process of re-tribalization? As shown above, early return-migrants such as Tama Huata, the son of Canon Wi Huata mentioned above, played a pivotal role in the creation and development of the Takitimu Rūnanga mo Te Reo which was, as mentioned, a direct precursor to the tribe's modern organisational structure Ngāti Kahungunu Iwi Incorporated. Return-migrants, however, continue to play important roles in the continued development of these tribal structures. One need look no further than Ngāti Kahungunu Iwi Incorporated itself for an example of the effects of return-migration as most of the permanent staff had spent the majority of their life living, working or being educated in major urban centres. In turn, this re-creation of modern tribal structures, and the associated renaissance in social and cultural values that it has brought about, has provided increased impetus for migrants to return home.

In order to shed light on these processes a number of interviews were conducted with return-migrants using local Māori organizations as a point of contact. In all, 42 such migrants were approached, and 33 of these agreed to be interviewed. These interviews were informal, often conducted in the interviewee's home, but structured around a set of questions regarding the factors stimulating return-migration.

Almost all Ngāti Kahungunu return-migrant interviewees advanced the view that a desire to be a part of the process of cultural and tribal regrowth occurring around the country (often termed "the Māori Renaissance") had played a major factor in their desire to return home. Common views were summarized with phrases such as "a desire to get in touch with my Māori side again" and "to allow my children to know where they come from" to emphasize this point. Interestingly, the majority of Ngāti Kahungunu return-migrants claimed that while a lack of employment opportunities in large urban centres had helped prompt their return home, the prospect of even higher rates of unemployment in the traditional tribal territory had not negatively affected their desire to return home.[21] This is interesting as it stresses the importance of cultural over economic factors in the Māori place–utility matrix. Another interesting demographic feature to emerge from interviews was the fact that those who returned were either the original rural–urban migrants of the immediate post-World War II era, or else descendants of those first migrants who were returning with parents or grandparents who had been part of the initial rural–urban migration. In every instance, returning families contained at least one member who had at one point lived in their "traditional" territory.

Many of those interviewed had brought with them a wide variety of skills acquired in the cities. This growing skills base is only now beginning to be utilized by the region's tribal organizations as they attempt to further solidify their organizational and economic structures. Accordingly, the re-creation of Ngāti Kahungunu tribal structures has had an impact, albeit small, on the regional labour market. Throughout the 1980s and 1990s, funds were increasingly made available to "Māori Service Providers" who would compete for contracts offered by government departments to provide services to the Māori community – particularly in regard to health services. While a number of these providers are private companies, a large number are tribally based – operating at either the Iwi or Hapū level, or through the new Iwi structures that have emerged over the last twenty years. A large number of the key figures behind these institutions are return-migrants who learnt their professional and managerial skills in their time away in the major cities of New Zealand and Australia. The existence of this purchaser–provider split in current New Zealand government policy has thus created another avenue through which modern Māori tribal structures may receive funding in order to further consolidate their existence.

Conclusion

The experiences of Ngāti Kahungunu serve to demonstrate the very real consequences that the interaction between changing Māori mobility patterns, government policies and Māori socio-political structures have wrought in New Zealand. The most recent manifestation of this is an emergent pattern of return migration to Iwi. However, one must be careful not to read too much into this dynamic. For one thing the numbers involved are still small, although as time goes on and the country's tribal structures continue to develop, more Māori may decide to return to their traditional tribal territories. Secondly, the fact that 26 per cent of all individuals who identified as being of Māori descent at the 1996 census gave no Iwi affiliation, greatly complicates the analysis of migration. No reason for this lack of Iwi identification was elicited by the census, so it is not possible to differentiate between individuals who were unable to identify with an Iwi, as opposed to those who were unwilling to identify with an Iwi. This lack of Iwi identification obviously makes return-migration difficult, if not impossible, to detect in many cases. Even though many Māori around New Zealand have responded to the opportunity to identify their Iwi affiliation through a free-phone Iwi helpline established by the Treaty of Waitangi Fisheries Commission – Te Ohu Kai Moana – many more still have no way of achieving this.[22]

The existence of a three-tier migration dynamic within Māori society – local-level migration, regional migration and international migration – should be noted. A number of the return-migrants that played a significant role in the re-creation of Ngāti Kahungunu's modern political structures spent considerable time living in urban centres in New Zealand, Australia, Europe and America before returning to their traditional homes. However, return migration should not be viewed as the new twenty-first century mobility dynamic for Māori. Rather, it is simply yet another example of the myriad of complex patterns of mobility found within the modern Māori population to have emerged from interactions between government policy and Māori socio-political structures. Nonetheless, while it would be an overstatement to claim that the return migration of Māori to their traditional tribal territories has led to the regrowth of the power base of Māori tribal institutions, it would not be an overstatement to argue that it has been a significant contributing factor. What this means for Māori at the dawn of the twenty-first century is unclear. All that we can be certain about at this point in time is that shifting patterns of Indigenous mobility in New Zealand continue to reflect the changing relationship over time and space between Indigenous peoples and their institutions, on one hand, and the changing social, economic and political needs of the settler-states that they now find themselves part of.

Notes

1 I would like to thank Maria Bargh, Bruce Buchan, Bethan Greener-Barcham, Toon van Meijl and John Taylor for constructive advice on successive drafts of this chapter. I would also like to thank Statistics New Zealand and the Centre for Aboriginal Economic Policy Research at the Australian National University for their financial assistance in obtaining the statistical data-sets used in the writing of this chapter. Needless to say, responsibility for any errors or omissions remain with the author.

2 Evidence would tend to suggest that this internal migration was at a lower level than that of the pre-contact period and in the early years of European contact.

3 Iwi experienced a short-lived resurgence in power with the recruitment of Māori soldiers for World War II along tribal lines. This resurgence however was short-lived as these returning soldiers settled at a variety of different locales around the country, including cities, in order to find work in the post-World War II era. This scattering of the servicemen prevented the regrowth of tribal sentiment gaining any political advantage.

4 The classic study of this phenomenon by Metge (1964) is still the best single piece of research on the Māori rural–urban migration of the mid-twentieth century.

5 This is not to say, however, that the New Zealand government's policy formulation along this new trajectory was always consistent. The 1953 passage of the Town and Country Planning Act, for example, severely restricted the ability of Māori to build homes on Māori land and so directly contributed to urban drift.

6 The Hunn Report argued that integration with broader New Zealand society should be the policy goal for Māori. The key idea behind the report was thus that Māori should adapt and conform to Western cultural norms and ideals. The report recommended "modernizing" Māori in order to solve social and economic problems with extra attention to be paid to the issue of education in an attempt to break the cycle of poverty that Māori were trapped within.

7 Interview with Peter MacGregor, Manager of Te Puni Kōkiri, Hastings Branch (Hastings, 16 January 2001).

8 The 1971 protest by members of Ngā Tamatoa – a group composed of young Māori intellectuals and professionals – against continued Government inaction regarding issues surrounding the Treaty of Waitangi was the crucial catalyst to the passage of this legislative act. The protest attracted so much public attention that the Labour government was forced to publicly approach the New Zealand Māori Council (NZMC) to help resolve these issues (Walker 1989: 163). The response by the NZMC then laid the basis for the legislation that was to end in the creation of the Waitangi Tribunal in 1975.

9 In 1977 Kara Puketapu was made Secretary of Māori Affairs, and in an attempt to develop a new philosophy of Māori development Puketapu launched the Tū Tangata strategy – a range of policies that sought to make a Māori world-view the basis of Māori development. The policies created under the auspices of this Tū Tangata strategy were to have dramatic effects in the 1980s.

10 Mātua Whāngai, Maccess and Mana Enterprises were all policies designed to empower Māori through the active involvement of Māori in their own development. Mātua Whāngai concerned the fostering of at risk Māori children, Maccess was a job skills programme and Mana Enterprises was designed to help Māori set up small businesses.

11 Elected in a political landslide in 1984, once in office the Labour government began to enact one of the most comprehensive periods of social and economic reform ever witnessed in the country's history. Based on a combination of ideologies including neo-classical economics and new public management the new government, once elected, proceeded to deregulate the economy at an unprecedented rate.

12 For an analysis of the reasons behind, and the consequences of, this urban–rural migration in the Western world see Jacob (1997), Jedrej and Nuttall (1996) and Jobes (2000). For an exposition on a similar dynamic operating in the general New Zealand population see Waldegrave (1998) and Hamilton (1999).

13 Before 1991 questions concerning Iwi affiliation were not included in census data and so it is not possible to discern whether or not these migrants were moving back to their traditional tribal areas or not.

14 A study by Rhema Vaithianthan (1995) on the relationship between Māori inter-regional migration, unemployment and Iwi affiliation found that Māori were less responsive than other ethnic groups to changes in regional employment conditions, due she theorized to their greater level of attachment to specific territorial areas. In summing up Vaithianathan concluded that Māori were willing to live in an area of high unemployment if that area happened to be their traditional tribal homeland (Vaithianthan 1995: 140–1).

15 For a more in-depth analysis of the history of Ngāti Kahungunu see Ballara (1991).

16 As an interesting aside to this migration pattern the majority of these Ngāti Kahungunu migrants were from the northern and southern sections of the tribe's territory due I would argue to the fact that those from the middle section were able to find jobs in the large freezing works located in and around the urban centres of Hastings and Napier in Hawkes Bay – an area located in the centre of Ngāti Kahungunu's tribal territory. Due to the lack of information on tribal affiliation in census before 1991 I am unable to obtain precise numbers of those who left, although anecdotal evidence from interviews and from publications of the time seem to indicate that this movement was quite substantial. This changed in the late 1980s as the closure of the two largest freezing works in the region, Whakatu and Tomoana, meant that many Ngāti Kahungunu became unemployed. Initial investigation would tend to suggest that many of these newly unemployed migrated to Australia in search of new

jobs. The long-term effects of this move are still to be investigated although the return of some of these international migrants in the late 1990s appears to be directly related to the increased incidence of the phenomenon of return-migration.

17 Interview with Peter MacGregor, Manager of Te Puni Kōkiri, Hastings Branch (Hastings, 16 January 2001).
18 Interview with Tama Huata (Hastings, Thursday 18 January 2001).
19 These were the local council regions created under the Māori Community Development Act 1962 which were discussed earlier.
20 For a general discussion of the interaction between neo-liberal ideals and the development of Māori policy over the last twenty or so years see van Meijl (1996). For an analysis of the consequence that the government's devolution policy had on Māori society see van Meijl (1997).
21 Thus when the first round of interviews were conducted with return-migrants for this chapter the major urban centres of Auckland and Wellington had unemployment rates of 15 and 16 per cent while Hawke's Bay's unemployment rate was running at just over 18 per cent.
22 The Treaty of Waitangi Fisheries Commission – Te Ohu Kai Moana (TOKM) – was set up in 1992 to hold fisheries assets returned to Māori by the Crown and to arrange for their eventual distribution. The settlement viewed the assets as being for all Māori via Iwi, but in the 1996 census around 143,000 people of Māori descent did not identify their Iwi. TOKM established the telephone service as a mechanism to help Māori identify their Iwi and ultimately benefit from the fisheries settlement. Those using the helpline had to provide any details they knew about their ancestors and the service attempted to identify their Iwi connection. The information was then sent to the caller and forwarded to the relevant Iwi organization, and it was up to those two parties to make contact. As of 30 September 2000 the Iwi helpline had received calls from 6,400 people (since its inception in November 1996), and of those callers positive identification was made for about 2,100 people.

References

Ballara, A. (1991) 'The Origins of Ngāti Kahungunu', unpublished PhD Thesis, Wellington: Victoria University of Wellington
Barcham, M. (1998) 'The challenge of urban Māori: reconciling conceptions of indigeneity and social change', *Asia Pacific Viewpoint*, 39(3): 303–14.
Barcham, M. (2000) '(De)constructing indigeneity', in D. Ivison, P. Patton and W. Sanders (eds) *Political Theory and the Rights of Indigenous Peoples*, Cambridge: Cambridge University Press.
Bedford, R. and Goodwin, J. (1997) *Migration and Urban Population Change: A Preliminary Analysis of the 1996 Census Data*, Briefing Paper Number 6, Hamilton: Population Studies Centre, University of Waikato.
Boddington, W. and Khawaja, M.A. (1993) 'Regional migration during 1986–91 with special emphasis on the Māori population', in D. Brown and C. Mako (eds) *Ethnicity and Gender: Population Trends and Policy Challenges in the 1990s*, Wellington: Population Association of New Zealand and Te Puni Kōkiri.
Boddington, W., Smeith, G. and Khawaja, M.A. (1993) 'A statistical analysis of ethnic and sex differentials in New Zealand mortality 1985–1987', in *Demographic Trends 1992*, Wellington: Statistics New Zealand.
Bu, J. (1993) 'Urbanisation and labour force transformation of the Māori

population 1956–1981', in D. Brown and C. Mako (eds) *Ethnicity and Gender: Population Trends and Policy Challenges in the 1990s*, Wellington: Population Association of New Zealand and Te Puni Kōkiri.

Butterworth, G.V. (1991a) *Nga Take i Neke Ai Te Māori: A Journeying People*, Wellington: Manatu Māori.

Butterworth, G.V. (1991b) *Nga Take i Neke Ai Te Māori: A Review of the Research*, Wellington: Manatu Māori.

Durie, M.H. (1998) *Te Mana Te Kawanatanga: The Politics of Māori Self-Determination*, Auckland: Oxford University Press.

Hamilton, E. (1999) 'Counterurbanisation and the rural idyll: a case study of lifestyle blocks in Dunedin', unpublished MA Thesis, Otago: University of Otago.

Hunn, J.K. (1960) *Report on Department of Māori Affairs*, Wellington: Government Printer.

Jacob, J. (1997) *New Pioneers: The Back-to-the-Land Movement and the Search for a Sustainable Future*, University Park: Pennsylvania State University Press.

Jedrej, C. and Nuttall, M. (1996) *White Settlers: The Impact of Rural Repopulation in Scotland*, Luxembourg: Harwood Academic Publishers.

Jobes, P.C. (2000) *Moving Nearer to Heaven: The Illusions and Disillusions of Migrants to Beautiful Places*, Westport: Praeger.

Maaka, R.C.A. (1997) 'The politics of diaspora', paper presented at the *Treaty of Waitangi: Māori Political Representation Future Challenges Conference*, New Zealand Centre for Conflict Resolution, the Institute for Advanced Legal Studies, and the Institute of Public Law, Wellington: University of Victoria.

Mabbett, D. (1988) 'Regional policy', in *Future Directions: The April Report of the Royal Commission on Social Policy, Vol. III, Part 1*, Wellington: Royal Commission on Social Policy.

Metge, J. (1964) *A New Māori Migration: Rural and Urban Relations in Northern New Zealand*, London: Athlone Press.

Ministry of Māori Affairs (1988a) *Tirohanga Rangapū: Partnership Perspectives*, Wellington: Ministry of Māori Affairs.

Ministry of Māori Affairs (1988b) *Te Urupare Rangapū: A Discussion Paper on Proposals for a New Partnership*, Wellington: Ministry of Māori Affairs.

Parr, A. (1988) 'Māori Internal Migration: 1976–1986', unpublished MSc Thesis, Canterbury: University of Canterbury.

Pool, I. (1991) *Te Iwi Māori: A New Zealand Population Past, Present and Projected*, Auckland: Auckland University Press.

Poulsen, M.F. and Johnston, R.J. (1973) 'Patterns of Māori migration', in R.J. Johnston (ed.) *Urbanisation in New Zealand: Geographical Essays*, Wellington: Reed Education.

Poulsen, M.F., Rowland, D.T. and Johnston, R.J. (1975) 'Patterns of Māori migration', in L.A. Kosinski and R.M. Prothero (eds) *People on the Move: Studies on Internal Migration*, London: Methuen.

Rose, N. (1996) 'The death of the social? Refiguring the territory of government', *Economy and Society*, 25(3): 327–56.

Rose, N. and Miller, P. (1992) 'Political power beyond the State: problematics of government', *British Journal of Sociology*, 43(2): 173–205.

Scott, K. and Kearns, R. (2000) 'Coming home: return migration by Maori to the Mangakahia Valley, Northland', *New Zealand Population Review* 26(2): 21–44.

Statistics New Zealand (1994) *New Zealand Now: Māori,* Wellington: New Zealand Statistics.

Swindell, K. and Ford, R.G. (1975) 'Places, migrants and organisations: some observations on population mobility', *Geografiska Annaler,* 57B: 68–76.

Te Puni Kōkiri (1998) *Māori Towards 2000,* Wellington: Te Puni Kōkiri.

Vaithianathan, R. (1995) 'The impact of regional unemployment and Iwi (Tribal) affiliation on internal migration', unpublished M.Comm. Thesis, Auckland: University of Auckland.

van Meijl, T. (1996) 'Redressing Maori grievances: the politics of compensation settlements in contemporary New Zealand', unpublished paper presented at the *European Society for Oceanists' Conference on Pacific Peoples in the Pacific Century: Society, Culture, Nature,* 13–15 December, Copenhagen.

van Meijl, T. (1997) 'The re-emergence of Maori chiefs: "devolution" as a strategy to maintain tribal authority', in G.M. White and L. Lindstrom (eds) *Chiefs Today: Traditional Pacific Leadership and the Postcolonial State,* Stanford: Stanford University Press.

Waldegrave, C. (1998) *Family Dynamics Among Urban to Rural Migrants: A Study of Households Moving Out of Urban Areas in New Zealand,* Wellington: Ministry of Agriculture and Forestry.

Walker, R. (1989) 'Māori identity', in D. Novitz and B. Wilmott (eds) *Culture and Identity in New Zealand,* Wellington: GP Books.

Walker, R. (1995) 'Māori people since 1950', in G.W. Rice (ed.) *The Oxford History of New Zealand* (2nd edition), Auckland: Oxford University Press.

9 American Indians and geographic mobility

Some parameters for public policy

C. Matthew Snipp

Frequent geographic mobility has been a tradition among many Indigenous societies dating back to their earliest origins. However, the coming of Europeans complicated those movements, and in many instances eliminated them altogether. For example, in the nineteenth century, nomadic North American Indians were confined to small areas of land reserved for their occupation. They did not win the right to leave these places until a United States Federal Court decided in the *Standing Bear* v. *Crook* case (1879) that such confinement was illegal. Over the years, American Indians, like many other Indigenous populations, have moved with ever-greater frequency between their reservation homes and urban areas. In modern times, the public policy implications attending the geographic movement of Native peoples have grown to a level of complexity unimaginable even in the late nineteenth century.

One reason for this complexity stems from the patterns of movement found among Native people. Different patterns of movement arise from different circumstances, and the antecedent circumstances associated with different types of mobility necessarily require different kinds of policy response to deal with them. For example, mobility that is short term and temporary may involve sets of migrants with needs that are different from migrants who are relocating permanently.

Another source of complexity concerns the government institutions responsible for developing and implementing policies that may either influence, or be influenced by, the geographic mobility of Indigenous populations. In the United States, for example, some movements of Native people demand responses from national or local levels of government, while other types of mobility are of special concern to the governing bodies of tribal communities. Add to this the administrative difficulties generated by overlapping jurisdictions and the need for inter-governmental coordination, and the task of targeting public services to many Native Americans becomes daunting, if not intractable.

This paper considers some of these issues with respect to how different patterns of migration entail special needs for Native people in the United

States, and in particular, how different groups of migrants have special needs with respect to public goods such as housing, education, employment, health care, and family services. Of course, many of these needs are not unique to Indigenous populations. For instance, a large influx of young families regardless of their ethnic background will tax local resources for education, housing, and family services such as day care. However, Indigenous populations pose additional considerations such as the preservation of ties with tribal communities, the maintenance of cultural traditions and practices, as well as issues related to their legal and political status in their home communities and in the nation at large.

The geographic mobility of Indigenous Americans

Three major types of movement are relevant to thinking about the public policy implications arising from the mobility of Indigenous populations. The first is the movement of Indigenous people from the countryside to cities. Not surprisingly, being settled out of the mainstream of American society has meant that American Indians, in particular, have been concentrated in remote rural areas. After World War II, American Indians began moving in sizable numbers to urban locations. Part of this movement was due to special programmes designed to encourage American Indians to move away from their reservation homes (Fixico 1986).[1] However, a large component of this migration has also been the result of post-war employment opportunities in cities at the same time that few opportunities existed on reservations (Bernstein 1991). In 1930, barely 10 per cent of all American Indians lived in urban areas. Fifty years later, about one-half of all American Indians could be found in cities (Snipp 1989).

While rural to urban migration has been the most common type of movement for Native Americans in the twentieth century, it is not the only type of movement found in this population. In recent years, migration from urban areas back to tribal communities in rural areas seems increasingly common. The observation that this type of mobility has increased in recent years rests heavily, if not exclusively, on anecdotal evidence. When data from the 2000 census become available, it should be possible to document with much more certainty and accuracy the extent to which this type of mobility is occurring. However, it is possible to speculate about the sources of this migration.

In particular, two developments that occurred in the 1990s may have precipitated a growing number of persons who moved back to reservations. One is the increase in the numbers of American Indians over the age of 65 who retire and return to their reservation communities. For many American Indians who moved to cities in the 1950s and 1960s, retirement and a pension income provide them an opportunity to return to the places where they were born and raised, and where they may still have many family members.

An even more dramatic development in the 1990s was the spread of casino gambling on many reservations. Gaming has a mixed record of success in Indian country. However, for reservations with even moderately successful casino operations, the income from these establishments forms the resource base for job creation in a variety of local economic sectors. Because most reservations invest their casino revenues in local developments, many of these communities have witnessed modest to spectacular job growth in their communities. As jobs have become more plentiful on reservations, this has caused American Indians who might have otherwise remained in cities to return to the reservation in search of employment. For example, the Menominee reservation in Wisconsin has sawmill and logging operations, in addition to a modestly successful casino/hotel establishment. In the 1990s, Menominee county was one of the fastest growing counties in the state.[2]

The temporal dimension of population mobility is the third consideration that has important implications for public policy. It is impossible to distinguish precisely between long-term and short-term migration. However, for the sake of discussion it is useful to think of short-term mobility as moves which take place within a year or so, very roughly approximating the seasonality that accompanies some types of employment (e.g. construction, farm labour) or school attendance. Long-term migration, again very approximately, is reflected in moves that take place not more than once about every five years. So, for example, a long-term rural–urban migrant is someone who moves to a city for an extended period – approximately five years or more for the sake of this discussion. Conversely, a short-term urban–rural might be someone temporarily settling in a rural reservation while waiting for a job to begin.

Short-term mobility is associated with certain types of occupations and with students attending college or boarding school. A key characteristic of short-term mobility is that it often tends to follow a circular pattern. For example, American Indians from reservations in upstate New York move to New York city for work in the construction trades. During slack periods when work is not available due to bad weather or economic downturns, these workers return to their reservations to reside with family or friends. Similarly, American Indian youths attending school may return to their reservations in the summer months when classes are not in session. Beyond anecdote, the prevalence of short-term mobility among American Indians is nearly impossible to document. However, a small amount of evidence indicates that such mobility is very widespread (Hackenburg and Wilson 1972; Weibel-Orlando 1991).

Geographic mobility, Indigenous peoples and public policy

There are a host of important public policy issues that are connected with the migration of Indigenous people. Again, many of these concerns are not unique for Native people. Urban newcomers for example, depending on their circumstances, may require employment, health, or housing assistance. The stress of relocation may lead to problems related to alcohol abuse, or result in domestic violence. Large numbers of urban newcomers may threaten to overwhelm the existing resources available for dealing with these problems and needs. Similarly, economically disadvantaged populations typically require more assistance in relocation, and may face greater problems becoming integrated into a new environment than persons with substantial material resources at their disposal.

None of these issues are unique for Native peoples. Migrants from a variety of ethnic backgrounds face these problems to a greater or lesser degree depending on the resources they possess. However, Indigenous populations are unique insofar as they include disproportionate numbers of economically disadvantaged persons for whom relocation involves considerable hardship. For example, the American Indian population is one of the least educated and most unemployed groups in the United States (Sandefur *et al.* 1996). For an American Indian with little education and limited job experience, the obstacles to becoming settled in an urban location are formidable. In particular, urban labour markets favour experienced well-educated workers. Without these qualities, American Indians coming to urban areas often find themselves jobless, and lacking the resources to obtain shelter in costly urban housing markets.

Besides these considerations, Native people also bring a special set of issues into play by virtue of their Indigenous status. One is the problem of preserving cultural traditions in settings alien to these practices, and often distant from the places where these traditions form a regular part of everyday life. For example, many American Indian groups have ceremonies and other observances that coincide with certain seasons and with family events such as births, deaths, and marriages.

At first glance, one may wonder how these traditions, even if important to preserve, have any connection with public policy. Yet, there is a connection when these observances run foul of local ordinances, such as those that govern the disposal of corpses, or even local sensibilities about what constitutes acceptable use of public areas such as park lands. As Native people seek to maintain their culture in alien environments, it is incumbent upon local officials to be aware of and sensitive to these practices and seek to accommodate them as well as they can. This accommodation clearly has public policy implications.

A second issue unique for Native people is most problematic for tribal governments, and more generally for Native political leaders.[3] This is the issue of maintaining connections with tribal people who move away from the reservation, reserve, or village. Again, at first glance one may wonder why this is an important matter for public policy. It is, in fact a very important problem insofar as Native people do not always permanently move away from their communities and desire to maintain their connections with friends and family. By the same token, community leaders must be concerned about meeting these wishes for two reasons.

One is that keeping these connections intact is a way of generating outside support when there are contentious problems impacting upon the community. This support may come in the form of money donations, letters to elected representatives, and other expressions of concern. For the reservation community itself, it is a way of appearing larger and with more resources than would be otherwise possible depending on local residents alone. Another reason is that Native migrants who leave their community are often the brightest, best educated, and most talented members of the reservation, and they leave because their reservation may offer few opportunities. It is especially important to maintain connections with these individuals in the events that jobs and other opportunities become available on reservations. Maintaining connections with these individuals, and facilitating their return to the community is one important way that Native localities can resuscitate and build up their human capital – an essential component of local economic development.

Some tribes go to extraordinary lengths to maintain these ties. For example, the Ho-Chunk in Wisconsin have established satellite offices to assist tribal members living in nearby urban areas. Others, such as the Cherokee nation in Oklahoma accommodate political participation through the use of absentee ballots. In recent years, the numbers of absentee ballots cast in Cherokee elections have become so large that some candidates for tribal office travel to distant locales such as Los Angeles to campaign where there are sizable numbers of potential Cherokee voters. In the 1999 election held for the office of Principal Chief, 48 per cent (6,137 out of a total of 12,756) of the ballots cast were absentee. It is impossible to determine the extent of involvement these absentee voters have in their local communities. Undoubtedly, some of these votes were cast by persons temporarily away from their communities for a short time. Others, however, may be away for many years and only occasionally (if ever) visit the communities impacted by their vote. In either case, the ability to cast an absentee ballot does provide a connection, however tenuous, with the traditional community and enables Cherokees to participate in local affairs, regardless of time or distance.

A third special concern for Indigenous communities is the social integration of persons who return to the community. Again, this may be less of a problem for outside authorities, such as the federal government, than it is for tribal leaders. Yet for local leadership, it may be a significant problem depending on the numbers of persons returning to a community.

The social integration of return migrants may pose the usual material concerns such as providing housing, jobs, schools and health facilities. This, in fact is not especially unique for Native communities. What is most different about many Native communities is that they are often very tightly knit social networks that do not always easily accommodate newcomers, even when they are long-time friends and family. This is not to suggest that Native communities are not warm and welcoming places. Rather that an influx of outsiders, especially in large numbers, cannot avoid being at least minimally disruptive at some level, and this disruption is not always a welcome event.

These disruptions may often be small, and not much of a concern for local policy makers. For example, it may involve inappropriate conduct at ceremonies and other events by persons less familiar with them, or it may involve much more serious problems such as the importation of behaviours and lifestyles that are at odds with traditional customs. For instance, reservations in the Plains and in the southwestern USA have been dealing with a rising number of their youth belonging to urban street gangs. Some speculation suggests that American Indian youth learn first-hand about gangs and gang behaviour living in places such as Los Angeles. Youths from urban areas returning to reservations have brought with them the attire and violent behaviour associated with street gangs, where it quickly spreads in popularity.

Finally, the maintenance (and lack thereof) of the distinct political and legal status claimed by Native peoples is a very unique problem for policy makers, and again, for Native political leaders. In the United States, American Indians who move to urban areas relinquish in most respects the rights of self-government they retain in reservation areas. Urban American Indians cannot claim legal jurisdiction over the land where they reside. Likewise, there is no precedent whatsoever for urban Indian organizations to be accorded the same rights of recognition that the federal government extends to tribal governments on reservations.

By the same token, state and local governments must assume the responsibility for providing services to these newcomers, duties that were once in the purview of federal and tribal government authorities. However, there is anecdotal evidence to suggest that local officials are reluctant to assume these responsibilities, and indeed may expect the federal government or even the tribal governments to remain a principal source of social support for these migrants.

In fact, there are precedents for these expectations. In the early 1930s the federal government established the Johnson-O'Malley Act to reimburse public schools for the education of American Indian children. Today, the United States Department of Education continues to offer special assistance to schools with significant numbers of American Indian children. Similarly, the Indian Health Service provides a limited amount of support for clinics that deliver health care services to urban American Indians. However, Indian Health Service assistance in urban areas is but a tiny fraction of the overall budget of this agency, and it continues to be focused on delivering care to reservation residents.

The lack of formal legal recognition for urban Indian groups, even formally incorporated organizations, creates a number of problems for them. One is that they cannot avail themselves of many of the services and sources of assistance that the federal government directs to reservations. For example, the single largest agency concerned solely with American Indian issues is, not surprisingly, the Bureau of Indian Affairs (BIA). Nonetheless, virtually all of the programmes administered by the BIA are directed exclusively to the needs and concerns of reservations. In fact, there is a great deal of opposition by reservation communities to diverting these scarce resources to urban communities. Given that BIA funded services are available only in the confines of reservation life, this obviously provides to persons who need or desire such services an incentive to move away from under-served urban areas.

Reservations, by virtue of having a land base and the ability to exercise jurisdiction over it have a form of capital unavailable to urban organizations. Some tribes have utilized this capital very effectively to establish gaming, tourism, agriculture and other sorts of enterprises (White 1990). However, urban Indian organizations seeking to expand economic opportunities for their communities must rely on conventional approaches to enterprise development: identifying entrepreneurial opportunities, securing loans with banks, and successfully competing with other businesses in the local market. Needless to say, the record of urban Indian organizations in economic development is virtually nonexistent compared to the efforts of reservations.

Lastly, not only do urban Indian groups lack recognition from the federal government, they also may have to vie for recognition from state and local governments. That is, state and local governments do not automatically recognize American Indians or American Indian organizations among the diverse constituencies they serve. Because the urbanization of American Indians is a relatively recent phenomenon, local government authorities are often simply unaware of the presence of American Indians in their community. As a result, American Indians and particularly American Indian organizations must make their presence in the community known to local officials. This was especially true in the 1960s and 1970s. However, newly formed organizations also must actively establish their legitimacy as representatives of American Indian interests within the plurality of interests that compete in city politics.

The implications of duration and place

As noted earlier, the public policy implications of Indigenous mobility vary according to the duration of migration, whether short- or long-term, and the origins and destinations of the movers, notably whether they are moving to or from a reservation or a city. The balance of this chapter looks at each of these types of movement in terms of the most common characteristics of movers, and the special needs they may have as migrants that pose special challenges for public and tribal officials.

Short-term rural–urban migrants

Among Indigenous people, short-term rural–urban movers are among the most common types of migrants (Snipp 1996: 45–6). In the United States, these are individuals who move from reservations to urban areas, most often in search of employment and educational opportunities. They tend to be relatively young, frequently single, and very often male. Their high mobility reflects a transitional stage in life – moving to find a first job, to change jobs, or to begin a programme of education or job training. This group presents public policy challenges to both urban officials and to leaders in tribal communities.

In urban communities, it is important to facilitate the adjustment of newcomers to an environment that in most respects is alien and hostile. Accordingly, in the 1960s and 1970s, a number of studies focused on the process of adjustment to urban environments by American Indians from rural reservations (Ablon 1964; Bahr *et al.* 1972; Hackenberg and Wilson 1972; Sorkin 1978). For many American Indians, city life was found to be a highly stressful experience that could lead to personal problems ranging from family discord to substance abuse.

Many Indian migrants from small isolated tribal communities scattered across the United States, have few if any friends or family living in the city to help with their transition. Unlike other ethnic minorities, American Indians who move to cities do not settle in established enclaves where they may have access to an extended support network to assist with their relocation. In the absence of an informal support network to rely on, American Indian migrants depend more on institutional sources of support such as public housing, welfare, and other social services. In many instances, responsibility for this support is assumed by the individual agencies and/or programmes in which the migrants participate. For example, many educational and job training programmes devote considerable attention to resolving personal difficulties associated with relocation. However, few resources for assistance are available outside of such formal programme structures.

For American Indians lacking well-established connections in urban locales, inter-tribal urban Indian Centres play a vital role in helping individuals and families become connected with networks of support that might assist them. Most urban Indian Centres were established during the 1960s and 1970s in cities with large American Indian populations. With support from the federal government, these centres provided job training, employment, and other social services. Though the funding for these centres has diminished significantly in recent years, they continue to serve as a focal point for urban American Indian communities. In this regard, they can help urban newcomers connect with other American Indians living in the area, as well as providing limited social services and a venue for cultural events such as powwows, feasts, and other gatherings (Weibel-Orlando 1991).

As a place for cultural events, Indian Centres have special appeal to short-term urban immigrants. Lacking other social connections, these migrants may have a special need and/or desire to be involved in a setting that bears some semblance of familiarity. Food, music and other activities provide to these migrants a connection with "home" that is almost certainly absent in the rest of their lives. In this respect, Indian centres can serve as a temporary "home" for short-term rural–urban migrants because they provide a place readily available for culturally expressive activities.

However, it is also important to understand that urban Indian Centres are typically located in areas where the American Indian population is relatively large. Cities such as Los Angeles, Oakland, Tulsa, and Minneapolis have one or more such centres. In other places where the Native American population is much smaller, Indian Centres are often nonexistent. American Indian migrants who move to such places must rely on existing services provided by local government, and are often confronted by service providers who are likely unaware, unfamiliar, or insensitive to Native culture and traditions.

Long-term rural–urban migrants

American Indians who settle more or less permanently in urban locales do so for a variety of reasons. The most common motivation for a long-term relocation is in connection with employment. This was especially common after World War II. Many American Indian veterans returning from service overseas found that life on the reservation offered them little but poverty and economic hardship (Bernstein 1991; Fixico 1986). Many of these American Indians chose instead to remain in urban settings where jobs and other opportunities were more plentiful. Likewise, well-qualified American Indians who participated in the federal relocation programmes of the 1950s, 1960s and 1970s also tended to more or less permanently settle in their new residences (Clinton *et al.* 1975).

Consequently, it should not be surprising that long-term rural–urban migrants include a significant number of persons who have been relatively successful in the job market. While it is true that urban American Indians experience a high rate of poverty and unemployment in urban areas, it is nonetheless true that urban American Indians, especially long-term relocatees, are generally better off than their rural reservation counterparts (Snipp and Sandefur 1988). Indeed, the economic opportunities offered by urban labour markets may be precisely the reason why these individuals have settled in urban areas on a long-term basis. In general, this group is more successful, more residentially stable and less at risk from developing the personal problems that can stem from the stresses of relocation. In light of these qualities, one might guess that long-term rural–urban migrants have few special needs in connection with public policies. Nonetheless, policy issues do arise in connection with these migrants, especially for tribal leaders.

At the outset, it should be said that while these migrants may have negotiated the move to an urban area with some success, they still face the same kinds of problems faced by other minority populations in urban areas. For example, discrimination in housing, employment, the courts, and other venues may be a serious problem. In some parts of the United States where American Indian tribes have asserted their rights to resources guaranteed by treaties, these initiatives have generated a substantial amount of anti-Indian backlash (Whaley and Bressette 1994). As a result, it is entirely possible that some of these sentiments are manifest in discriminatory actions by landlords, employers, and the police.

Instead of being a group with special needs for services, long-term rural–urban migrants also might be viewed as a resource for policy makers and public officials. In particular, this group is an especially important resource for reaching out to recently arrived urban immigrants to aid them in the transition to urban living. They are familiar with the difficulties of making the transition from reservation to city life, such as those already mentioned. They are likely knowledgeable about the lifestyles and cultural practices that recent immigrants have left behind and this obviously facilitates communication. And lastly, long-term immigrants have the kinds of skills and information required for successfully relocating to an urban setting. This information may be as simple as where to shop for certain kinds of goods or where to find inexpensive housing. In these ways, long-term immigrants play an important role in helping to resettle newcomers, and possess specialized knowledge that is likely unavailable in conventional public programming.

While long-term rural–urban movers may have a special value to city officials, they pose a special dilemma to leaders in tribal communities. In particular, the extended duration they spend away from their tribal community may raise a reasonable doubt about whether they should be rightfully considered a member of these communities. Quite simply, if an individual leaves the tribal community and does not return for, say twenty years, is it reasonable for that community to continue recognizing that person as someone who has a legitimate right to participate in community affairs such as elections for tribal council?

Many tribes such as the Cherokee of Oklahoma allow all tribal members, regardless of their residence to participate in such events. However, assuming that good government depends on an informed electorate, one may wonder about the quality of voter decisions when they reside hundreds or even thousands of miles from their home communities. Furthermore, it is also easy to see how such practices can lead to tensions between those residing in the tribal community and those who participate from afar.

For example, members of tribal communities may believe that these connections should not, or need not, be maintained. Such sentiments may be fuelled by traditional values that underscore the obligations of individuals to maintain some sort of connection to their home communities, no matter how tenuous (Brant 1999). If individuals who leave the reservation fail in this obligation, this can be interpreted as a violation of honoured tradition by those left behind. For example, many tribes have ceremonial events that occur throughout the year. It is expected that tribal members, if they are truly members of the community regardless of their current place of residence, will make whatever efforts are necessary to return to the reservation to fulfil family or clan obligations and expectations regarding their participation in tribal events. Thus, any long-standing absence is likely to be interpreted as a severing of ties to the community. Needless to say, in these circumstances, persons returning to the reservation after an extended hiatus in contact, can attract the same suspicion and distrust afforded to strangers and outsiders.

Short-term urban–rural migrants

In many respects, persons who move back to their reservation communities for short spells may be similar to, if indeed not the same, persons who move to urban areas for brief periods of time. However, there is one very important difference and this has less to do with the migrants themselves and much more with the policy officials who must be concerned with these individuals. That is, to the extent that they require services and assistance from the community, they are the concern of tribal and federal authorities. By virtue of their reservation location, urban–rural migrants are no longer the subject of concern for state and local officials, nor are they candidates for programmes and policies administered by these authorities.

Because short-term urban–rural migrants likely share many characteristics in common with their short-term migrant counterparts in urban areas, they have many of the same needs. On the other hand, their reasons for returning to the reservation may be different and therefore have different implications for local policies and programmes. For example, taking a job or attending school have traditionally been common reasons for relocating to an urban area. On the other hand, individuals returning to a reservation community (at least until recently) typically have not moved for employment or school-related purposes. If anything, a return to the reservation very likely signals some other sort of crisis. At the very least it indicates an issue of sufficient gravity to precipitate a significant change in lifestyle. In some cases a return to the reservation may be to care for an elderly parent or a sick relative. In other cases, it may be caused by a job loss and inability to find new work. In many, if not most instances, urban–rural migration is a response to some sort of personal situation likely unrelated to material gain, and this quality differentiates it from migration in the reverse direction.

The temporary quality that attends this type of migration may incline local leaders and programme administrators to see the needs of these individuals as problems that will resolve themselves. That is because these migrants are by definition individuals temporarily settled on the reservation and, as such, whatever special needs they may present are also a temporary problem. However, when the numbers of these individuals are sufficiently great, they may represent a type of community underclass that cannot easily be ignored (Sandefur 1989). In particular, short-term housing and employment may prove to be especially important concerns for this population.

On many, if not most reservations housing is a scarce commodity. Housing is not only scarce on reservations, it also tends to be substandard and overcrowded (US Department of Housing and Urban Development 1996). Consequently, for individuals returning to their tribal communities, finding suitable housing may be as difficult as finding a place to live in the city. The lack of housing for short-term migrants is an acute problem for tribal leaders for at least two reasons.

One is that the usual strategy for dealing with a lack of reservation housing is simply to call upon the generosity and good will of friends and family. Data do not exist to document the frequency with which this occurs but anecdotally, there is good reason to believe it is fairly common. However, it is also well known that reservations have some of the most overcrowded housing stocks in the United States (Snipp 1989; US Department of Housing and Urban Development 1996). The discomfort, stress, and occasional lack of safety that overcrowded housing represents is only exacerbated by the arrival of newcomers seeking lodging, however temporary.

Policy makers concerned with housing may regard this as a relatively small, and indeed transitory problem that flares intermittently on their reservation. For example, it is most problematic during economic down-turns and seasons when construction and similar sorts of work are unavailable. However, there is another more immediate problem this creates in connection with the administration of public housing: namely, most public housing projects limit the number of occupants who may reside within a single unit, and exceeding this limit may be grounds for eviction. Short-term migrants pose a potentially significant problem when their presence results in routine violations of occupancy limits. Policy makers, and especially the administrators of public housing, need to be attuned and sensitive to the realities of reservation housing shortages and the need for a flexible approach to enforcing regulations at the local level.

Employment for short-term migrants may also be an especially acute problem for this population. Under any circumstance, employment is often scarce on many reservations. Compared with long-term residents, short-term migrants face additional limitations. One is that the need to find work may be an urgent one. For example, an individual who moves home to take care of some kind of family need requires a ready source of income that could not be anticipated in advance of the move. When a move is planned and anticipated, part of the planning may be finding a job in advance of relocating. For short-term movers, their moves may be unexpected and made without plans for employment.

Short-term migrants may also face a disadvantage finding employment if it becomes known to prospective employers that an applicant does not intend to remain on the job for more than a short duration. In instances where a job is temporary or of short duration, this is not a problem. However, to the extent that employers prefer to hire permanent workers, this may limit the employment prospects for short-term movers. Ideally, reservation tribal governments would be equipped to assist short-term migrants seeking work. This could be done with temporary placements in tribal offices or businesses, but in reality there are few if any such measures to be found in Indian country.

Long-term urban–rural migrants

The presence of long-term urban–rural migrants in reservation communities is a relatively recent phenomenon, and still a relatively uncommon one. Nonetheless, urban–rural migration, particularly to reservations has become more commonplace in recent years. There are two reasons for this development.

Before World War II, American Indians were heavily concentrated in rural locations, usually reservations. However, as noted earlier, this changed dramatically after the war as secular trends and federal programmes fuelled migration to urban areas. By 1980, about as many American Indians lived in cities as in rural settings (Snipp 1989). Many of these urban migrants moved to cities to find work, and as many of them are now reaching retirement age, the reservation is for many an appealing place. It is appealing for a variety of reasons: proximity to kin, attachment to childhood memories, access to healthcare from tribal or Indian Health Service clinics, and relatively inexpensive housing. Thus, one can see a small but growing number of American Indian retirees returning to the reservations where they were raised (John 1996; Weibel-Orlando 1988).

A second precipitating factor related to urban–rural migration is the job growth and economic opportunities that have developed on many reservations. It is true that many reservations still have chronically depressed economies, but some have been able to develop successful agriculture, forestry, fishing, manufacturing, or tourist industries. Casino gambling has produced some of the most spectacular and widely noticed success, but it is not the only source of jobs on many reservations today which include those in service industries. As a result of job growth, American Indians who wish to return to their communities now have opportunities that have been heretofore nonexistent. The upshot has been an increase in the numbers of persons returning to reservation communities.

As noted above, tribal leaders must find ways to maintain strong connections with those tribal members who leave their communities. Clearly, long-term movers are the most serious challenge in this regard. For the reasons already explained, the nature of the relationship that tribal communities maintain with members who live away for lengthy periods of time is complicated. Tribal communities may wish to maintain these connections because these distantly settled tribal members may be a source of economic or political support. They also may represent a source of scarce expertise in the event they can be persuaded to return to their origins.

For example, one way of attracting legal or business expertise would be to make sure tribal members in possession of law or business degrees feel welcome to return when they retire from the regular workforce, or sooner if so desired. In particular, well-educated retirees and others with specialized skills who feel motivated to offer them to their tribal communities are an important resource that tribal communities may wish to cultivate.

Alternatively, while some tribal members who reside away from their communities for long periods may be seen as a resource if they return, there are others who may engender costs to the community, real or perceived, upon their return. For example, one potentially unwelcome return migrant might be someone who has lived away from the reservation for years but decides to return to take advantage of housing, job opportunities, or medical care available to tribal members. Socially integrating such individuals back into the community is made doubly difficult by the likely perceptions that these individuals have moved back only to take advantage of tribal resources. And while these individuals may be fully entitled to these resources, such perceptions are almost certain to spark resentment and even conflict in the local community (Anders 1999). In recent years, as some reservations have become increasingly wealthy from casinos, tourism, and other sorts of economic development projects, this has fuelled debates over whether tribal members who remain away from the community for long periods are still connected to tribal traditions (Trosper 1999), and thus entitled to tribal resources.

The policy challenge

Relatively high rates of population mobility are a defining characteristic of Indigenous American populations. The nature of this mobility is such that significant challenges arise for policy makers who are concerned with servicing the needs of Indigenous groups. In the United States context, those concerned with formulating such policy include the leaders of tribal governments as well as their non-Native counterparts outside the reservations.

Given the many different reasons for population movement, and its many different outcomes, no single policy prescription can address the heterogeneity of circumstances associated with those moving. However, there are some broad parameters that policy makers should recognize when considering the needs of migrant populations. Specifically, these are concerned with sustaining and enhancing the cultural, political, and economic viability of Indigenous communities.

The preceding discussion has focused on the necessity of finding ways to transport and reproduce Native culture through space and time. As Indigenous populations become more mobile, there can be little doubt that they risk, literally, leaving their culture behind. The greater the distance from, and time away from their traditional communities, the greater the risk that migration can rend the cultural fabric. Similarly, the maintenance of the special legal and political statuses of American Indians, is made problematic by the movements and consequent dispersal of community members.

In the United States, this special legal status is all but completely surrendered by those who choose to leave their reservation communities. By the same token, reservation communities that wish to remain recognized as viable entities by the federal government require some small number of permanent residents. The Bureau of Indian Affairs does not have a threshold number below which tribal recognition is rescinded, although a number of small reservations were disbanded in the 1950s. On the other hand, to carry out the functions normally associated with tribal governments requires some minimum, albeit unspecified number of residents, to justify the presence of full-time fire and police protection, and other governmental services. On small reservations where a full complement of public services is prohibitively expensive given the small number of persons utilizing them, the Bureau of Indian Affairs may resort to consolidating services across several small reservations, or contracting with other units of local (non-Indian) government (towns, counties, school districts) to provide services. This is not uncommon in areas such as California where there are many small reservations. In fact, the Johnson-O'Malley Act enacted by the US Congress was specifically concerned with providing funds so that Indian children could attend local public schools.

Thus, population mobility has undeniable consequences for those who leave their Native communities, as well as for those who remain. For these reasons, finding ways to maintain social ties with emigrating tribal members as well as finding ways to reintegrate returning migrants is of utmost importance for tribal communities.

Within the context of these issues, which are unique to the Indigenous component of the United States population, there is also a broad spectrum of concerns that are common to all migrant populations. Access to housing, employment and education, along with the emotional strains of moving to an unfamiliar locale, to name only a few, are among the many issues that confront Indigenous migrants. Policy makers who plan social programmes, and local officials who administer them, must recognize that Native populations have higher than average mobility, and at the same time appreciate that the relatively small numbers of Native persons in many urban settings may cause them to be virtually invisible as a group.

For the foreseeable future, Native Americans are likely to remain a highly mobile population given their need to access urban opportunities and the difficulties they experience in securing residential stability. Thus, the policy issues outlined in this chapter are likely to be salient for the next decade or longer. The challenge to tribal leaders and public policy makers alike is to find innovative and creative ways of addressing these matters in ways that recognize the distinct character of Native Americans. The formulation of public policy shaped in ways that recognize these distinctive characteristics, would represent an enormously important shift in direction compared with past practices.

Notes

1 In the early 1950s, the United States Congress passed a series of laws intended to abolish reservations and re-settle the Indian population in a handful of pre-selected cities. This legislation and the programmes established by it are collectively known as "Termination and Relocation." The purpose of these policies was to hasten the assimilation of American Indians by severing their ties to the reservation and integrating them into urban labour markets. By most assessments, these programmes were a dismal failure.
2 Annual population estimates for the Menominee reservation are not available. However, such estimates are available for Menominee county, which is almost entirely covered by the Menominee reservation. Menominee county consists of the Menominee reservation plus a small tract of land that is adjacent to, but does not belong to the reservation. The county land adjacent to the reservation includes a small settlement of vacation homes. However, most of the growth taking place in Menominee county is taking place on the reservation, which includes a newly constructed (1995) settlement built with tribal funds to accommodate tribal members returning to the reservation.
3 In the United States, tribal governments are democratically elected political entities that have governing jurisdiction over the territory occupied by the tribe. Their authority is subordinate to the federal government but with few exceptions, they are not subordinate to other jurisdictions such as state and local governments. Like other governmental entities, tribal governments have legislative powers, the power to police and enforce tribal codes, and provide a range of social services such as health care, education, housing assistance, and public safety.

References

Ablon, J. (1964) 'Relocated American Indians in the San Francisco Bay Area: social interactions and Indian identity,' *Human Organization*, 23: 296–304.

Anders, G.C. (1999) 'Indian gaming: financial and regulatory issues,' in T.R. Johnson (ed.) *Contemporary Native American Political Issues*, Walnut Creek: Altamira.

Bahr, H.M., Chadwick, B.A., and Day, R.C. (1972) *Native Americans Today: Sociological Perspectives*, New York: Harper and Row.

Bernstein, A. (1991) *American Indians and World War II*, Norman: University of Oklahoma Press.

Brant B. (1999) 'The Good Red Road: journey's of homecoming in Native women's writing,' in D. Champagne (ed.) *Contemporary Native American Cultural Issues*, Walnut Creek: Altamira.

Clinton, L., Chadwick, B.A. and Bahr, H.M. (1975) 'Urban relocation reconsidered: antecedents of employment among Indian males,' *Rural Sociology*, 40: 117–33.

Fixico, D.L. (1986) *Termination and Relocation: Federal Indian Policy, 1945–1960*, Albuquerque: University of New Mexico Press.

Hackenburg R.A. and Wilson, R.C. (1972) 'Reluctant emigrants, the role of migration in Papago Indian adaptation,' *Human Organization*, 31: 171–86.

John, R. (1996) 'Demography of American Indian elders: social, economic, and health status,' in G.D. Sandefur, R. Rindfuss, and B. Cohen (eds) *Changing Numbers, Changing Needs: American Indian Demography and Public Health*, Washington DC: National Academy.

Sandefur, G.D. (1989) 'American Indian reservations: the first underclass areas?' *Focus*, 12: 37–41.

Sandefur, G.D., Rindfuss, R.R. and Cohen, B. (1996) *Changing Numbers, Changing Needs: American Indian Demography and Public Health*, Washington DC: National Academy.

Snipp, C.M. (1989) *American Indians: The First of This Land*, New York: Russell Sage.

Snipp, C.M. (1996) 'The size and distribution of the American Indian population: fertility, mortality, residence and migration,' in G.D. Sandefur, R. Rindfuss, and B. Cohen (eds), *Changing Numbers, Changing Needs: American Indian Demography and Public Health*, Washington DC: National Academy.

Snipp, C.M. and Sandefur, G.D. (1988) 'Earnings of American Indians and Alaska Natives: the effects of residence and migration,' *Social Forces*, 66: 994–1008.

Sorkin, A.L. (1978) *The Urban American Indian*, Lexington: Lexington Books.

Trosper, R.L. (1999) 'Traditional American Indian economic policy,' in D. Champagne (ed.) *Contemporary Native American Cultural Issues*, Walnut Creek: Altamira.

US Department of Housing and Urban Development (1996) *Assessment of American Indian Housing Needs and Programs: Final Report*, HUD-1574–PDR (May), Washington DC: US Department of Housing and Urban Development.

Weibel-Orlando, J. (1988) 'Indians, ethnicity as a resource, and aging: you can go home again,' *Journal of Cross-Cultural Gerontology*, 3: 323–48.

Weibel-Orlando, J. (1991) *Indian Country, L.A.: Maintaining Ethnic Community in Complex Society*. Urbana-Champagne: University of Illinois Press.

Whaley, R. and Bressette, W. (1994) *Walleye Warriors: An Effective Alliance Against Racism and for the Earth*, Philadelphia: New Society Publishers.

White, R.H. (1990) *Tribal Assets: The Rebirth of Native America*, New York: Henry Holt and Company.

10 The formation of contemporary Aboriginal settlement patterns in Australia

Government policies and programmes

Alan Gray

"The European settlers wanted land", according to Berndt and Berndt (1988: 497). With this observation they succinctly introduce a discussion of the "dispossession and depopulation" (Elkin 1951) of Aboriginal peoples that extended throughout most of the two hundred years after the first European settlement in Australia in 1788. The fact that the land was already occupied before the establishment of British colonies was not recognized in Australian law until the Mabo High Court decision of 1992 (Reynolds 1994). Smith (1980a: 193) characterizes Aboriginal settlement before white invasion as highly mobile but also restricted in geographical scope. The high mobility of the pre-contact era had been voluntary, for the management and sustainable exploitation of the natural resources that fell within the tribal boundaries of each Aboriginal group. White intrusion through pastoral expansion, the gold rushes and agricultural intensification treated prior Aboriginal occupation of the land as a mere inconvenience, and eventually led to involuntary resettlement of the Aboriginal survivors in supervised institutions (Smith 1980a: 195). The aim of this chapter is to review how control over Aboriginal mobility was achieved and enforced, and how this control was used to achieve less overt aims of responsible government agencies, then to examine by detailed example the process by which recent Aboriginal migration and settlement patterns developed after the breakdown and elimination of the mechanisms of control in the latter decades of the twentieth century.

Control over mobility

Many writers have described how Aborigines were subject to dispossession, pacification by force, protection and assimilation, in the words of Elkin (1964: 363), or displacement and decimation, coalescence and concentration, dispersal and diffusion, in the words of Smith (1980a: 194). It is not necessary to repeat the detail of unequal warfare, massacre, rape, casual murder and theft of land that eventually led to the concentration of remnant Aboriginal populations in reserves and missions. For present purposes, it is sufficient to note that the single central aspect of government policies and programmes for Aborigines was, at the very beginning, to control mobility by removing Aboriginal people from their land and confining them in concentration camps.

The first attempts at doing this, with the establishment of Black Town (in present-day Sydney) and other initiatives by Governor Lachlan Macquarie in 1814, have been seen as a response to Aboriginal needs (Smith 1980a: 196), although the same writer refers to a "dominant, darker tradition" which rapidly succeeded the original initiative. Indeed, for most of the nineteenth century it was the establishment of Christian missions rather than government activities that served settler purposes. The exception was Tasmania, where genocide commenced in the first year of settlement in 1804 with a massacre at Risdon, and continued without let for the first 25 years of the colony (Berndt and Berndt 1988: 505; Bonwick 1870; Smith 1980b: 170), until finally the remnant Aboriginal population of 159 persons was concentrated on Flinders Island in Bass Strait between 1830 and 1833. The fact that government intervention to "protect" Aborigines came relatively much later, compared with the period of genocidal dispossession, was to be repeated in all the mainland States and the Northern Territory.

For example, while in the State of Victoria there were some pre-existing Christian missions and one recently opened government-run settlement, or station, already in operation, a Protection Board charged with operating concentration institutions was not established until 1860 (Long 1970: 15), and this was the first of the mainland States to do so. Apart from the lapsed attempt at Blacktown in New South Wales, further moves towards the establishment of government stations did not occur until the appointment of a Protector of Aborigines in 1881 (Long 1970: 26) and the establishment of the Aborigines Protection Board in the following year (Rowley 1970: 10), although there had been many previous missionary activities organized under a non-government organization called the Aborigines Protection Association. The term "protectorate" had actually been used referring to Aborigines in the title of a New South Wales parliamentary report as early as 1849 (New South Wales Parliament Legislative Council 1849). All of this makes it very clear that the colonial governments had been entirely satisfied to leave the issue of the protection of Aborigines in missionary hands until the situation had evolved beyond dispossession to a point where it would be uncontroversial for government to move beyond a laissez-faire approach and become actively involved in concentration.

While the purpose of missions, government stations and reserves in all States was to remove Aboriginal people for the convenience of white settlement, the managerial control exercised over the inhabitants of these institutions constituted the central factor in the existence of Aboriginal people for as long as the institutions existed. Middleton (1977: 66) provides a useful summary of the extent of the powers vested in the managers, whether missionary or bureaucratic:

> Mission and government settlement superintendents were granted magisterial and other powers and administered laws controlling employment, Aboriginal marriages, miscegenation, maintenance of children, care of minors, education, compulsory action in cases of leprosy, venereal and some other diseases, the supply and consumption of alcohol, possession of firearms, the removal of Aboriginal "camps" near towns, the enforced transfer of people to and from reserves, control of property, and the suppression of so-called "injurious customs" (with missionaries quite free to decide which traditional religious and social customs fitted into this category).

In the more remote areas of central, northern and western Australia, there were still Aborigines living independently of such missions and government settlements until the 1950s and 1960s, but even in the remotest areas the programme of concentration eventually resulted in most Aborigines being relocated into managed settlements by the end of the 1960s, apart from communities located on cattle stations.

The primary purpose of concentration camps was always to remove Aborigines for the convenience of non-Aboriginal land use, whatever the rhetoric. For example, on Cape York Peninsula in northeast Australia, the Aboriginal community at Mapoon was simply shifted to a completely different area for the convenience of aluminium ore mining during the 1960s. Elsewhere in the tropical north, Middleton (1977: 155–6) records the case of apparently benign Commonwealth (federal) Government assistance to the community at Yirrkala in 1970 to take their land rights case, over the mining area at the Gove peninsula in Arnhem Land, to the Supreme Court. The case failed because the court perceived that the relationship of the people with the land was religious, and not economic ownership. Middleton argued that this perception was actually assisted by expert evidence that emphasized the religious aspects of relationship to land, rather than the economic aspects that had been severed by expropriation for mining. (See Figure 10.1 for localities referred to in the text.)

Concentration camps could be economically successful, but the successful ones were also subject to greedy eyes that wanted the land. A number of examples are cited by Rowley (1970: 67). The station that had been established in Victoria before the Protection Board was inaugurated in 1860, was initiated by Aborigines as a successful farming enterprise (Gray 2001). After the Board was established, the residents of this and the other eight stations that came into existence were moved from one station to another, as each was closed or opened, until in the end there was only one station left, at Lake Tyers in the east of Victoria (Long 1970: 16). There

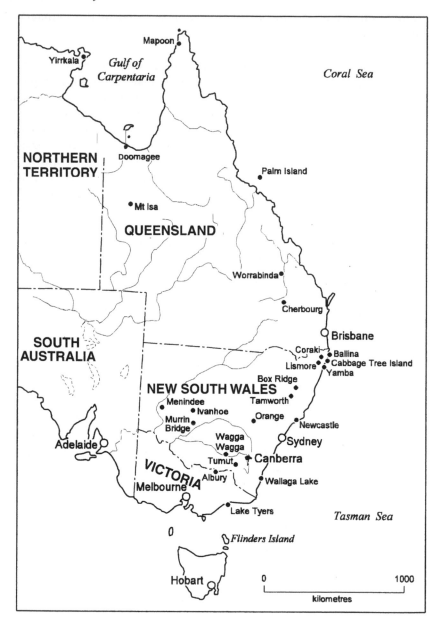

Figure 10.1 Localities referred to in the text.

should be no doubt that this process of moving people from one place to
another represents a further phase in the administration of dispossession,
about which rather little has been written, although it has been one of the
major features of the policies followed in all States. It is dispersal.

Dispersal

The term "dispersal" has been used by many writers in a rather general way to describe the process by which tribal groups were severed from their land and split into clusters of people who somehow ended up in different places. In this chapter "dispersal" is used in a much more specific sense, to describe the deliberate activities of missions and governments to ensure that attachment to land and tribal group was so completely eliminated that there could be no future claims to land on grounds of traditional owner- ship. This process was followed assiduously by governments in all parts of eastern and southern Australia up to the 1970s. It was begun also in the north, but because it was carried out there bureaucratically rather than expressed in policy statements, it is barely documented. To bureaucrats working for the Protection and Welfare Boards, dispersal in this sense would have been regarded simply as part of the practice of assimilation, without need for any more extensive justification.

The practice of dispersal was clearly of interest to writers like Long (1970), who visited and surveyed government institutions in Queensland, New South Wales and Victoria in 1965, because Long made a point of recording what he could discover about the origins of the residents. For example, the earliest-established surviving Aboriginal settlement in New South Wales, which had been established at Wallaga Lake in 1891, contained people from distant parts of the State in 1965, and Long notes that residents had established connections with people from the Lake Tyers reserve in Victoria through exchanges and family links (Long 1970: 62). It is also the case, from personal observation, that some families were removed from Wallaga Lake to other locations in New South Wales. In this way, the location of the Wallaga lake settlement near the well-known sacred site at Mumbulla Mountain did not guarantee the existence of traditional close connection of all the residents to the site. On the other hand, there remained many residents who did have such a connection.

Because dispersal was not written policy, it was carried out in various different ways in New South Wales. A different approach was taken in the case of Murrin Bridge, established in the west of New South Wales in 1949. The people who were moved there had originally been concentrated at a location called Carowra Tank further to the west near the town of Ivanhoe, had then been moved away for the convenience of railway construction to the settlement of Menindee on the Darling River in 1934, and then to the Murrin Bridge location (Long 1970: 82; Rowley 1970: 67). Later, some people from easterly communities were moved there, but most Murrin Bridge residents still had Wongaibon origins in the west. It was just that the people had been relocated in a completely different place.

As a counter-example, consider the community at Cabbage Tree Island south of the town of Ballina on the north coast of New South Wales, which is a case of people still occupying part of their original land. The island was a retreat for Bundjalung people from gun-carrying horsemen who had carried out bloody massacres in the area during the nineteenth century. It became a government station in 1893, and like some other communities on the north coast the residents remained descendants of the original inhabitants. In the south and west of the State of New South Wales, concentration and dispersal were far more evident in the residence patterns of the station populations located there.

In Queensland, Long (1970: 102–28) and Anderson (1986) recorded how the stations at Cherbourg, Woorabinda and Palm Island would concentrate people from all over the State, including different Indigenous ethnicities from the Torres Strait, side by side with people from many other tribal or geographic origins. In Queensland, dispersal in that form was evidently more completely carried out than in New South Wales. Palm Island, in particular, had a reputation as being the place to send any "trouble-makers" who were ready to question managerial authority or assert human rights, although other settlements were also used on occasion for the same purpose. In Queensland, the authorities seemed to have discovered that dispersal could be an instrument of authoritarian control as well as a means to eliminate traditional ties.

In the more remote parts of Queensland, in the far northwest and on Cape York, the reserve and mission communities were more likely to contain people who had originated in the area, although people originating from these groups could also be found in the southeastern communities mentioned above (Anderson 1986). For example, at Doomadgee north of Mount Isa there were Ganggalida people from the Gulf further north, Garawa people from across the Northern Territory border, and some Wanyi people and other small groups from the northwest. In the late 1970s, at a time when the mission was under control of Brethren missionaries, the Ganggalida people had been moved well away from the original site of the mission, which was within their own country. Still, they and the other major groups retained affiliation with their country and the Garawa were later successful in land claims in the Northern Territory. Of note here, as at Wallaga Lake in New South Wales, is the practice of dispersing people across State boundaries. In the Doomadgee case, concentration and dispersal had not been effective enough to achieve the desired effect of severing traditional ties, but the process had started and the mission superintendent there at the time was quite open about the desirability of breaking down tradition.

Gale (1967, 1972, 1987) has provided a comprehensive picture of the way in which Aboriginal people in South Australia were thoroughly redistributed, mainly, at first, through the activities of missionary groups who concentrated Aboriginal groups then shifted whole communities as they aggregated them into larger settlements. She notes how some of these aggregated communities became staging points for migration into the Adelaide metropolitan area.

An example of quite a recent programme of dispersal is the "Aboriginal Family Voluntary Resettlement Scheme", which operated in New South Wales in the 1960s to encourage Aboriginal families to move away from remote areas in the west of the State to resettlement sites in the urban areas of Newcastle, Albury, Wagga Wagga, Orange and Tamworth (Ball 1985; Mitchell and Cawte 1977; Morgan 1999). The description by these authors is useful because of their attempt to rebut unnamed critics of the scheme who had claimed that Aborigines were being moved away from their homes and encouraged to assimilate, leading to denial of identity. Mitchell and Cawte (1977: 30) state that "obviously the project leads Aborigines to embrace a way of life which revolves around a cash economy, but this is a universal trend and is hardly stemmed by an Aboriginal remaining destitute in an outback reserve where living conditions are deplorable". The rebuttal begs the evident question – why not improve the living conditions and develop economic opportunities at point of origin?

It is interesting that the writers had hoped to demonstrate improved physical and mental health among Aboriginal families who participated in the scheme, but the results they report are mixed. In fact, at one point of the paper they admit that not only was there no significant improvement in physical health, but "if anything, the position has apparently worsened" (Mitchell and Cawte 1977: 32). The scheme was an indirect government programme, because although it was operated by a non-government organization, it received Commonwealth funds and State staffing assistance to carry out its work. The guiding principles were stated in noble terms rather than the language of dispersal and assimilation (Mitchell and Cawte 1977: 29–30):

(i)   the migrant must initiate the move himself;
(ii)   sponsorship should be given to family units rather than single people;
(iii) regular intensive counselling will be required;
(iv)   housing and employment should be prearranged for intending migrants;
(v)   a sense of identity should be encouraged;
(vi)   psychological stress should be watched and minimized;
(vii) migrants who are unable to cope are to be assisted to return home without prejudice.

As a further hint of the true purpose of the scheme, it may be noted that all of the resettlement areas were places with few previous Aboriginal inhabitants.

Population dispersal activities had the same aims as the much more explicit activities which resulted in the so-called "stolen generations" of lighter-skinned children, who were removed from their parents in all States and Territories and raised institutionally, very often in ignorance of their true origins. The story of these stolen generations has now been told (Commonwealth of Australia 1997).

Urbanization

In the period after World War II, the administration of control over the missions and government stations started to lose its way, at least in south-eastern Australia. Long (1970: 31) pinpoints one reason for this occurrence. Using statistics from the Aborigines Welfare Board of New South Wales, he found that in 1940, 64 per cent of the "able-bodied" on reserves were in employment, but this rose sharply to 96 per cent by 1944, as a result of the labour shortages created by the absence of non-Aboriginal men, and some Aborigines, in the armed forces overseas. All those men in employment were working off the reserves. Also, the population of the reserves declined during the same period, a result of families working far away. The labour shortage on the reserves themselves sent agricultural activities, at that time seen as a means towards self-sufficiency, into sharp decline.

The agricultural activities never did recover properly, and the aim of self-sufficiency was abandoned during the immediate post-war years (Long 1970: 32). This sent the administration of the reserves in New South Wales into a period of aimless drift. Control over residence was still nominally in existence, but it was progressively loosened. While many country towns imposed restrictions which effectively prevented Aboriginal families from taking up residence inside the town boundaries, and Aboriginal children were kept out of the schools and out of the "Olympic" swimming pools that were built in many towns during the 1950s, gradually these restrictions were modified and relaxed and exceptions were made. Many older Aboriginal people can remember the petty restrictions and the humiliation they could cause. In Queensland and Western Australia, clearing the towns of Aborigines who had made their way there was practised as a "routine chore" of the police, regularly over many years, according to Rowley (1970: 121). Although Rowley was referring to it as a past practice at the time he was writing, it is doubtful that the practice had ended entirely.

For many Aboriginal people, the major metropolitan areas were easier migration destinations than nearby country towns. Mitchell and Cawte (1977: 34) cite census figures showing that in 1947, 7.6 per cent of Aborigines in New South Wales lived in Sydney, but by 1966 this had risen to 16.2 per cent. Expressed in terms of odds, in 1947 there was one Aborigine living in Sydney for every twelve living in the bush, but by 1966 the odds had been reduced from 12:1 to 5:1. Since the 1947 figure would have included people from the reserve at La Perouse within the metro-politan area, it is evident that during this period there must have been substantial migration into Sydney from rural areas of New South Wales.

A very detailed analysis of the early stages of migration to a major urban destination was undertaken for the case of Adelaide and South Australia by Gale (1967, 1972). The process of urbanization was also investigated by Smith (1980a) who was the first to observe that analysis based on figures of migration to the cities was incomplete without considering migration in the opposite direction, and he also lamented the absence of information on the age distribution of migration.

Both of these gaps in the data were filled by an analysis of the 1981–1986 period using census data (Gray 1989). This analysis found that, far from gaining Aboriginal population through migration, the largest metropolitan areas of Sydney and Melbourne were by the 1980s experiencing a net loss of Aboriginal population through migration. The analysis revealed a high degree of circular movement, and particularly a pattern of young unmarried adolescents and adults moving to the city to return to the country at older ages, married with children. This pattern was found in all major urban areas, but in cities such as Canberra, Darwin, Perth and Adelaide the net migration remained clearly towards the city. This was attributed tentatively to more active State housing programmes in those cities than in the southeastern capitals, and affordability.

Similar analyses were carried out subsequently by Taylor and Bell (1996, 1999) for the 1986–1991 and 1991–1996 periods, but unfortunately without the explanatory power obtained from examination of the age patterns of migration. In both periods, Sydney and Melbourne continued to experience net loss of Indigenous population through migration, while all other capital cities experienced Indigenous population gain. Taylor and Bell (1999: 20–2) find some similarities in the migration patterns with those of the non-Indigenous population but higher turnover in the Indigenous population.

While these patterns have now been verified over a period of fifteen years, there remains a common misperception that the Indigenous populations of the major metropolitan areas are increasing as a result of migration. On evidence and on logical grounds, there was a time in the past when migration to the cities of Sydney and Melbourne was occurring at a high level. On the same grounds of logical deduction, we can understand full well that while young people may be attracted to the big cities for employment, opportunities, and adventure, the costs of living in the largest metropolitan areas must be a disincentive to permanent settlement, especially when individuals get married and have children. A particularly important factor might well be the availability of suitable housing, a theme that will be explored in the remainder of this chapter.

It is important to record that the data that were used to reach the conclusions set out for the major cities are essentially unaffected by the problems of different levels of enumeration in different censuses, that have been discussed extensively by Gray (1997) and Taylor (1997). This is because each analysis uses data from one census only, namely responses to questions concerning location five years previously compared with current residence. Quite different conclusions can be reached by comparing numbers recorded in urban areas in successive censuses; these comparisons are seriously affected by different levels of enumeration, and are consequently misleading.

The data would essentially be no more difficult to interpret when dealing with urbanization in non-metropolitan regional cities and small country towns, except that the analysis is extremely complicated and has not yet been undertaken. As with the major urban areas, comparison of data about individual country towns using different censuses would be highly misleading. Accordingly, understanding of migration processes could be improved considerably by analysis of more geographically disaggregated data.

As Taylor (2001) has noted, urbanization is both an individual and a community-level process. A person "urbanizes" when he or she moves to an urban area, but net migration to urban areas is also described as urbanization, and the growth of small communities into towns and cities is another form as ever-spreading boundaries of existing urban areas urbanize the fringes. In demography and population geography, the term "urbanization" is usually reserved for growth of urban areas through net migration.

Historically, Smith (1980a) notes two important ways that Aborigines have become urban, referring to "New Black Towns" and "Black New Towns". The first reference is to Governor Macquarie's establishment of a new community for Aborigines away from Sydney Town at Black Town, now Blacktown in the southwest part of the Sydney metropolitan area. The second reference is to the occurrence of concentrations of in-migrants in central city areas like South Sydney, containing the suburbs of Redfern, Newtown and Alexandria (Anderson 1993). A "New Black Town" is therefore any community established as an Aboriginal settlement that develops into a township. Many examples have occurred as former institutions have developed into service centres for more scattered habitation in north and central Australia. A "Black New Town" is an urban concentration of migrants.

In fact, considerable numbers of Aborigines have become urban through two other processes. The first is movement into urban areas that does not include staging through in-migrant areas. This type of movement is typically into State rental housing, or sometimes into Aboriginal community-owned housing within urban areas, and is no doubt the most common form that Aboriginal urbanization takes, in any sense of the term urbanization. The second type is the process whereby some country towns in all parts of Australia have gradually come to have Aboriginal majorities in their populations. In the past, Aboriginal majorities could not deliver Aboriginal participation in local government activities, because in many parts of Australia voting rights for local government depended on property ownership. With the introduction of universal suffrage in Western Australia in 1985, the last remnant of restrictive voting rights disappeared, and although there were still practical problems in achieving effective Aboriginal participation in local government (Commonwealth of Australia 1992: 47–56; Sanders 1995), local councils and hospital boards increasingly reflect population composition. Favourable urban living conditions for Aborigines are favourable towards further urban population growth.

Housing and mobility

The key role identified for State rental housing in shaping Aboriginal urbanization has already been mentioned. Because Aboriginal families are mainly renters wherever they live (64 per cent of all Indigenous households were in rented dwellings in 1996), it is imperative to consider housing arrangements, affordability and availability in any detailed analysis of long-term migration patterns among Aboriginal people. Generally speaking, individual Aborigines and Aboriginal families do not have the economic power to be able to migrate by "selling up" in a rural location and buying straight into housing in an urban location. During at least the first twenty-five years after the Commonwealth Government involved itself in Aboriginal housing programmes, it is fair to say that it was these programmes that generated most new Aboriginal housing in Australia, although independently-managed State housing programmes directed by Aboriginal participation in management have increasingly supplemented the Commonwealth programmes.

Historically, Commonwealth Government programmes for Aboriginal housing had three distinct lines of emphasis, of which the most prominent, in terms of dollars committed, was the provision of rental housing through State housing authorities. The second line of expenditure was the provision of rental housing through Aboriginal community organizations, and the third was the extension of loans to Aboriginal families to purchase homes. These three types of housing programme delivered products that were essentially very different. State housing provided a roof over a family's head, for which rent was paid, and created a capital asset belonging to the State or Territory Government. Similarly, housing through an Aboriginal organization provided a home for which rent was paid by the family using the dwelling, but in this case the capital asset was the Aboriginal community's. Loans to Aboriginal families created capital assets for these families, subject to repayment of the loan.

These differences between the three types of programmes were far from superficial, yet it was often the case that the "product", Aboriginal housing, was seen as a meaningful entity, to the extent that the total number of houses provided through the three lines of expenditure were regarded as a meaningful programme output measure. If we take the trouble to view the three types of product in some detail from a consumer's point of view, then we can begin to appreciate that there are serious conceptual problems involved in grouping them together, especially in terms of understanding the links between the administration of housing supply and Aboriginal migration.

Take the case of State housing authority dwellings. Any family, Aboriginal or non-Aboriginal or both, could apply for State rental housing subject to whatever conditions applied in the State or Territory concerned. The difference in the case of Aboriginal families is that the Commonwealth's grants created a stock of dwellings to which Aboriginal families had priority access. This special housing provision was designed so that Aboriginal families might not need to wait as long for State housing as other families. The value for Aboriginal families was essentially the amenity value of obtaining State houses and flats more quickly than by waiting on the normal list for State housing. Acceptance of this conclusion leads to consideration of some uncomfortable consequences. If the waiting list for Aboriginal State housing was generally approximately as long as the waiting list for other State housing, in at least some States many Aboriginal families put their names on both lists and took whichever came up first. It is therefore a sustainable argument that special State housing for Aborigines was essentially a non-specific supplement to total programmes of expenditure by the State Housing Commissions and Departments, but with the benefit that houses were built more frequently in places to meet Aboriginal people's demand, for example in country centres, and there was a discernible impact on the awareness of Aboriginal housing needs within State bureaucracies.

Originally, most of the houses built under these programmes were standard State authority dwellings, in all senses – design, location, tenancy arrangements – except that they were earmarked for Aboriginal occupancy. Increasingly, there were ventures by the housing authorities into programmes of construction of housing in Aboriginal communities. Where programmes became specific in this way, they competed directly with the programmes of the Commonwealth's main Indigenous bureaucracy – the Aboriginal and Torres Strait Islander Commission (ATSIC) – which provided grants for rental housing to local community organizations. From the point of view of the tenant, there was little if any difference – especially where the State housing agency provided dwellings for administration by a local community organization.

The second line of Commonwealth expenditure on rental housing – direct grants to Aboriginal community organizations – created tangible assets from an Aboriginal point of view. Aboriginal people, through membership of the title-holding companies, had a common stake in ownership and a common interest in the allocation of houses to families within their communities. The programme of grants developed to include allocations for renovation of existing dwellings, special programmes for services to town camps[1] as well as for conventional housing, and related infrastructure services.

In a sense, because both types of expenditure discussed so far created rental housing, the difference between them might not be substantial from the point of view of a family paying rent. However, Aboriginal families in State housing were expected to conform to strict rules governing just who may live in a house (and pay more rent if their household included people outside the approved categories), and they could expect to lose the house if they failed to keep up with rent payments or other requirements. The application of rules in Aboriginal community-owned housing was based to a much greater extent on community decisions and rules which might be flexible enough to make allowances for the particular situation that a tenant might face. The size of a local community agency operating rental housing creates flexibility in assessing priorities and needs, to the benefit of the community; on the other hand, small size usually does entail limitation of options as well.

Unless Aboriginal organizations used grant money to build or buy houses in towns rather than on Aboriginal land, the assets that were created by this type of expenditure were almost totally unrealizable. Two of the factors working towards freezing these assets on paper were prejudice in the market and the inaccessibility or remoteness of many Aboriginal communities, but these were not the important issues. Even if a buyer could be found for a house in an Aboriginal community, the land on which the house was located was likely to be community-owned under legally binding arrangements of communal tenure which may forbid sale and would usually make sale impractical. On the other hand, it was possible to observe that Aboriginal organizations which owned houses in some urban areas were able to operate in entrepreneurial fashion to improve their assets by selling houses to tenants who could obtain ATSIC loans, and then purchasing other houses. In this case, the assets created by Government expenditure are more fluid.

While grants to Aboriginal organizations for housing comprised a readily comprehensible category of expenditure viewed from the Government side, the comments just made demonstrate that even within this category there were quite different types of housing viewed from the Aboriginal community's side. This is even more the case when it is remembered that grants might have included expenditure on infrastructure and maintenance. In Aboriginal communities in remote areas, there was scope for balancing housing programmes which provided new housing against the wider needs of such communities for management and improvement of existing facilities and shelter according to the community's priorities. The connecting thread between grant programmes was the creation of assets for community organizations and their ownership and management by those organizations.

One very important aspect of the programme of grants to Aboriginal organizations was that it was the only line of Commonwealth expenditure which had the potential to make improvements to old and inadequate housing in Aboriginal communities outside urban areas, with the exceptions which have been noted. State rental housing for Aborigines had an urban bias, and housing purchased through loans was exclusively urban.

The placement of a standard three-bedroom house in a small remotely located community could be entirely inappropriate to the needs of the community or any one of its members. On the other hand, the same community might benefit greatly from provision of some less specialized buildings for community use or use as core houses to be improved more gradually, from some community planning assistance and environmental improvements, and from ongoing assistance to maintain building stock. In Aboriginal communities in all parts of Australia, the newest, best and largest houses are the ones which have, on average, the largest number of inhabitants, as is sensible but does not necessarily suit design. Replacement of old hardly habitable houses with newer ones very often means that the older buildings will be immediately occupied by former community inhabitants anxious to return, especially in eastern Australia where population pressure in communities has for decades caused overflow to the nearest towns.

Houses that are not designed for local climatic conditions are not appropriate houses. Houses that are not designed with their likely use patterns foremost in mind are not appropriate houses. Houses that incorporate relatively sophisticated facilities for cooking, food storage, laundry and personal hygiene purposes are not appropriate if built where maintenance of these facilities cannot be delivered easily. Houses that, to operate efficiently, require reliable water and electricity reticulation are not appropriate unless their construction is accompanied by provision of this other infrastructure, including environmentally sound methods of sewage disposal. Many rural and remote Aboriginal communities are in locations where provision of these community infrastructure services is a major problem.

Aboriginal housing created by loans was again a completely different type of "product". Houses bought with such loans were almost always standard housing in towns and cities, far from rural communities. The houses became personal assets, eventually paid for by the occupants, but they were only available to families which could meet requirements enabling them to repay the loans. The eligibility criteria were well within the capacity of many Aboriginal families to meet, where relatively few would qualify for standard loans from banks and building societies, the loans had concessional arrangements about interest rates and repayments, and they included provision for home establishment costs.

The existence of a range of options, in the form of a range of very different housing "products" under Commonwealth and State programmes, could superficially be taken to be a sign that appropriate choices could be made to suit various categories of need. In fact, for most Aboriginal people the choices were extremely limited and not necessarily appropriate to their needs.

For example, with some exceptions the only programme which operated in Aboriginal communities outside the urban areas was the programme of grants to Aboriginal community organizations to provide housing and related services. A community-level approach worked on the assumption that the community's organization could represent, prioritize and articulate the different needs of families which made up the community. The operation of programmes in this way was consistent with aims of self-management for Aboriginal communities; access to housing, in this case, was a two-stage process in which families accessed community organizations to express their needs and the organizations accessed government agencies in an attempt to have the needs fulfilled.

Aboriginal people living in urban areas or moving to urban areas might appear to have had a range of choices, such as the private rental market, purchase of homes through bank and building society mortgages, purchase of homes with ATSIC loans, rental from Aboriginal community organizations (where they operated in urban areas), and rental from State housing authorities. For most families the choices were much more limited. Racial discrimination was an important factor in the private rental market because it was difficult to prove that it had occurred. Discriminatory practices confined Aboriginal families largely to down-market neighbourhoods and long-stay caravan parks in many urban areas. Mortgages, which constituted the next two options, were not available to Aboriginal people who were unemployed or had unsteady employment or received low incomes. The options were thus narrowed.

Those that are now left from the initial list are rental from Aboriginal organizations and rental from State housing authorities. Because Aboriginal organizations operated only in some urban areas and often then only in some parts of those towns or cities, the option of rental from State housing authorities was the only real choice available for many. The actual large proportions of Aboriginal families utilizing this type of housing was the evidence. Access to this type of housing, except in emergency cases, meant joining long waiting lists first. During the waiting period, families often had to live with other families, sometimes in breach of tenancy rules, or chanced the private rental market.

It should now be clear that migration to urban areas has had to conform with access to housing that was determined by Commonwealth and State government programmes. The final section of this chapter provides a detailed case study of how mobility was determined in this way in an area in northern New South Wales during the mid-1980s.

Case study: mobility and opportunity

The Aboriginal Family Demography Study was conducted by researchers from the Demography Department at the Australian National University between May and December 1986 in the far north coast area of New South Wales. The Aboriginal people there were mostly descended from the original inhabitants, who were speakers of the various dialects of the Bundjalung language. By the late twentieth century, few surviving people had been born before Aborigines in the area had been settled into the concentration camps called "missions" in New South Wales whether they were administered by churches or not. The last free Gidabal dialect people in the mountainous forests between Kyogle and the Queensland border had been forced into the mission at Muli Muli near Woodenbong (Ngudjumbuny) in the second decade of the century.

The study was based on multiple interviews in 100 households containing 544 people, living in Ballina, Box Ridge, Cabbage Tree Island, Coraki and Lismore. There was an emphasis in the study on housing arrangements. It was at the time still possible to see that the existing housing arrangements had been determined in the first instance by a period of construction of improved housing on the missions administered by the Aborigines Welfare Board, in the 1950s and 1960s. New Houses were built at Cabbage Tree Island in the 1950s (*Dawn* 1955) and at Box Ridge and Gundurimba Road in the early 1960s (*Dawn* 1964a, b). This programme eliminated the makeshift dwellings which had served until that time. The Cubawee mission was closed down, with many of the residents being moved to Gundurimba Road just outside Lismore, and others into homes within Lismore, and at Exton, Alstonville and Evans Head. In its mouthpiece newsletter (*Dawn* 1964c), the Aborigines Welfare Board was optimistic about the long-term effects ("in 30 years time") of their rehousing programmes, here referring to the nearby community at Yamba:

> The long view is often the better one. These little children who take their places day by day in school with their fellow pupils, should now have an incentive to learn. These mothers who dress their little ones so sweetly, should now have cause for pride, when they return home. These husbands who often get into the doldrums, should now see the joy of steady work.

This house-building activity was still the period of exclusion of Aborigines from the towns, still remembered by older Aborigines. The simple weatherboard houses built in that period or before, at Cabbage Tree Island, Box Ridge and Gundurimba Road, still constituted the majority of the houses rented by Aboriginal community organizations to Aboriginal families on the former missions during the 1980s. They were supplemented by a small number of newer houses, all built within the previous ten years. Of twelve households in the study that had occupied their houses continuously for more than twenty years, all but two were in these three locations.

After the Aborigines Welfare Board went out of existence when the Commonwealth Government took responsibility for Aboriginal affairs in New South Wales, a process completed in 1969, the houses that had been owned by the Board in urban areas were for a time administered by the Aboriginal Land Trust which had been set up to be responsible for former reserve land. The Trust, however, transferred these urban houses to the Housing Commission, the New South Wales Government body responsible for all State-owned rental housing. In the 1986 survey there were only two householders in Housing Commission houses who had occupied their houses for more than fifteen years, and both of these had last moved around 1969. Apart from these two, it therefore seems likely that the Board's urban houses were mainly lost from the stock of Aboriginal housing. In the early days of State housing earmarked for Aboriginal use, Aboriginal families were to be given the opportunity eventually to purchase their homes, but this promise had been withdrawn by the time of the study. Families which had improved their houses in the expectation that they would be able to purchase were angered by this change of policy.

As population growth strained the capacity of housing on the missions, the overflow was accommodated in the towns. While this generalization ignores the existence of exceptional families whose move to town was for other reasons, the reason mentioned most often by survey respondents for moving house was crowding in their previous accommodation. From 88 householders who were interviewed about their reasons for moving, 39 mentioned crowding or related reasons, including 17 out of 24 households in State housing. Many householders gave multiple reasons for moving, and 142 reasons were enumerated among the 88 households. After crowding, the next most commonly mentioned reason for moving was the need of the householders for their own homes, mentioned by 20 respondents.

There were very few references to dissatisfaction with the location, structure or standard of previous housing, and among these the most prominent was movement away from flood-prone areas, with three cases. In other words, there was nothing much wrong with the locations where they had lived previously, according to the householders. Also relatively infrequent were reasons related to costs or tenancy rules. Reasons referring to educational opportunities were uncommon and reasons referring to employment were totally absent. This is in stark contrast to the answers given in a national survey of internal migration among all Australians in the same year (Australian Bureau of Statistics 1987), where the most frequently stated reason for moving in the previous twelve months was for employment, and another important reason was to move closer to work or school.

Among the factors pushing families from where they had lived previously, crowding was overwhelmingly dominant, while the most prominent factor pulling families into their new housing was the need for homes of their own. These were both related to population pressure. When the answers given by householders were summarized as "push" factors and "pull" factors, it was found that 32 respondents stated their reasons for moving only in terms of attractive features of their new accommodation ("pull" factors), 25 gave reasons only in terms of unattractive features of their previous accommodation ("push" factors), and 19 gave a mixture of push and pull factors, while 12 householders did not give any reason that could be classified either way. To express the reason for moving only in terms of push factors might signal a lack of feeling of control over the process.

The pattern of movement from the missions into the towns could be demonstrated by analysing the places from which the residents lived before moving into the accommodation in which they were found. This information was not given in eight of the household interviews, and a further eleven households had moved into the study area, mainly from nearby parts of the far north coast area or from further down the coast. Of the remaining 69 households, 51 were outside the missions and at least 12 of these had moved from the missions the last time that they had moved house. The figure is not precise because the former locations were not always stated precisely. While three households had also moved in the reverse direction from town to mission, the weight of movement was clearly in the direction of the towns, and it was mainly into State housing in the first instance. Of the twelve householders who moved into the towns, ten went into State housing.

Most of the town dwellers who were not in State housing had already been living in the towns before their last moves. In 34 of these households, 29 of the householders had previously lived in one or other of the three urban areas included in the study (Lismore, Ballina and Coraki), three had come from outside, and only two had come from the missions. Homebuyers were a particularly interesting sub-group; seven were interviewed, and five of them had moved from rental housing in the urban areas, while one had been living with parents, and the other at a mission.

The stages of movement into the towns are reflected in the lengths of time that householders had occupied their dwellings. On average, the occupants of houses on the missions had occupied them for 14.5 years. By contrast, householders in State housing had an average length of occupancy of 7.0 years, the tenants of community housing in the urban areas had an average length of occupancy of 5.0 years, and the average for home buyers was also 5.0 years. The shortest average periods of occupancy were for households in caravans or privately rented dwellings, less than one year for both categories.

More recent movements of households demonstrate that this pattern of redistribution was continuing. Householders on the missions constituted 34 of the 88 housing interviews (39 per cent), but only 12 of the 46 householders who had moved in the previous five years (26 per cent), and a still smaller proportion, four from 22 or 18 per cent, of householders who had moved in the previous one-year period. In the last group of householders who had moved in the previous twelve months, five were in State housing, four were in caravans, four in privately rented dwellings, three were home-buyers and two were in rental housing owned by community organizations in the towns. These 22 households, it can be noted, constituted one-quarter of the 88 householders who were interviewed about housing.

Putting these pieces of evidence together, we can conclude that the arrangement of Aboriginal housing in the areas covered by the study resulted from a process with three clearly identifiable stages or steps. The first step was population pressure in the missions pushing families off into State rental housing in the towns. Next, there was considerable movement within the towns into other forms of rental accommodation, including caravan parks and dwellings rented privately on a short-term basis, but largely into housing rented out by Aboriginal community organizations. The study area was certainly not typical of country New South Wales in this respect, because the scale of community organization involvement in urban housing was greater than in most other places at that time. Thirdly, the small proportion of families who borrowed money to buy houses had nearly all come from rental housing within the towns.

The stages were continuing in the period immediately before the study; in other words the process was not separated in time but families were found at the three different stages of the process. This description was nevertheless specific to the second half of the 1980s, because it is evident that in later periods more and more urban-origin families would undergo their own transitional processes, while fifteen or more years previously the only movement into the towns had been into Aborigines Welfare Board rental accommodation. There were also major differences in the process in the three towns in the study area, for in Ballina there was no community-owned housing, and in Coraki the community housing was, typically, occupied directly after movement from the Box Ridge mission because the community organization built all its new housing in the town and none on the mission.

Yet it is possible to see in the overall pattern that something like this process must have happened no matter how government Aboriginal housing programmes operated. Population pressure has a grinding inevitability. What government programmes determined was how (and to what extent) the population overflow was accommodated in the towns.

Conclusion

The historical process of asserting heavy control over Aboriginal mobility stands in stark opposition to the hunting and gathering system of land use that it supplanted after initial periods of conflict and dispossession. The key elements of control were authoritarian institutional concentration and dispersal. The term "dispersal" is used to describe the deliberate activities of missions and governments to ensure that traditional attachments were eliminated.

As with any characterization, there were exceptions, one of which (pastoral station populations) has only been mentioned briefly in this chapter, and another (fringe-dwellers and town campers) has not been discussed at all. The general picture also varied considerably in the different States and Territories, and at local level within States. These exceptions and variations do not alter the general validity of the process which largely determined how Aboriginal settlement patterns had developed by the middle of the twentieth century or shortly after. Concentration was achieved everywhere, but dispersal was much more advanced in the south and east of the continent than in more remote regions.

During the last fifty years, the Aboriginal population has become much more urbanized after the breakdown and elimination of the mechanisms of residential control. The major point that has been made in this chapter is that residential patterns have changed in a way determined to a very great extent by the housing programmes operated by the Commonwealth and State Governments, because permanent or long-term migration requires access to housing.

The heavy emphasis on provision of State rental housing during the early period of Commonwealth housing programmes for Aborigines was found to be a key element in shaping urban residential patterns in a study conducted in northern New South Wales in 1986. Speculative conclusions have been drawn previously about the role played by provision of State rental housing in continuing migration inflows into the smaller capital cities, and the questions still remain for more detailed analysis.

In the case study, it was clear that one of the main reasons for urbanization of the Aboriginal population of the study area was simply population pressure in the former reserve communities. We noted in passing that the result was a diaspora, of people linked with any of these communities, to be found in the surrounding towns. This created a paradox where provision of improved housing on any of these communities did not necessarily result in decrepit housing being taken out of service, because it would immediately be occupied by people moving back from nearby towns.

The case study represented an evolutionary stage in the development of contemporary residential patterns, where larger and larger proportions of urban-dwelling Indigenous people are urban-origin Indigenous people. However, as the overall residential patterns unfold, observations about the processes of migration to urban areas, as set out in this analysis, are likely to remain valid as long as the economic status of Indigenous Australians sets them apart as renters of houses, not owners.

Notes

1 Town camps refer to discrete official or unofficial Aboriginal living areas within an urban area that have different arrangements from the town for municipal services, or no such services at all. For further insight see Sanders (1984).

References

Anderson, C. (1986) 'Queensland Aboriginal peoples today,' in J.H. Holmes (ed.) *Queensland: A Geographical Interpretation*, Brisbane: Queensland Geographical Journal 4th series, vol 1: 298–320.
Anderson, K. (1993) 'Place narratives and the origins of inner Sydney's Aboriginal settlement, 1972–73,' *Journal of Historical Geography*, 19(3): 314–35.

Australian Bureau of Statistics (ABS) (1987) *Internal Migration: 12 months ended 31 May 1986*, Catalogue No. 3408.0, Canberra: Australian Bureau of Statistics.

Ball, R.E. (1985) 'The economic situation of Aborigines in Newcastle, 1982,' *Australian Aboriginal Studies*, 1985/1: 2–21.

Berndt, R.M. and Berndt, C.H. (1988) *The World of the First Australians. Aboriginal Traditional Life: Past and Present* (4th edition), Canberra: Aboriginal Studies Press.

Bonwick, J. (1870) *The Last of the Tasmanians, or, The Black War of Van Diemen's Land*, London: Sampson Low.

Commonwealth of Australia (1992) *Mainly Urban: Report of the Inquiry Into the Needs of Urban Dwelling Aboriginal and Torres Strait Islander People*, House of Representatives Standing Committee on Aboriginal and Torres Strait Islander Affairs, Canberra: Australian Government Publishing Service.

Commonwealth of Australia (1997) *Bringing Them Home: Report of the National Inquiry into the Separation of Aboriginal and Torres Strait Islander Children from their Families*, Sydney: Human Rights and Equal Opportunity Commission.

Dawn (1955) 'Cabbage Tree Island Station: twelve new buildings now,' *Dawn*, 4(3): 6.

Dawn (1964a) 'The end of an era and the start of a new life at Lismore,' *Dawn*, 13(5): 12–13.

Dawn (1964b) 'Energetic north coast building program by Aborigines Welfare Board,' *Dawn*, 13(11): 4–5.

Dawn (1964c) 'Out of the night for Aborigines at Yamba,' *Dawn*, 13(8): 4–5.

Elkin, A.P. (1951) 'Reaction and interaction: a food gathering people and European settlement in Australia,' *American Anthropologist*, 53(2): 164–86.

Elkin, A.P. (1964) *The Australian Aborigines: How to Understand Them*, Sydney: Angus and Robertson.

Gale, F. (1967) 'Patterns of post-European Aboriginal migration in South Australia,' *Proceedings of the Royal Geographical Society of South Australia*, 67: 21–37.

Gale, F. (1972) *Urban Aborigines*, Canberra: Australian National University Press.

Gale, F. (1987) 'Aborigines and Europeans,' in D.N. Jeans (ed.) *Australia, a Geography, Volume 2: Space and Society*, Sydney: Sydney University Press.

Gray, A. (1989) 'Aboriginal migration to the cities,' *Journal of the Australian Population Association*, 6(2): 122–44.

Gray, A. (1997) 'The explosion of Aboriginality: components of indigenous population growth, 1991–1996', *CAEPR Discussion Paper No. 142*, Canberra: Centre for Aboriginal Economic Policy Research, The Australian National University.

Gray, A. (2001) 'Indigenous Australians: demographic and social history,' in J. Jupp (ed.) *The Australian People: An Encyclopedia of the Nation, Its People and Their Origins*, Cambridge: Cambridge University Press.

Long, J.P.M. (1970) *Aboriginal Settlements: a Survey of Institutional Communities in Eastern Australia*, Canberra: Australian National University Press.

Middleton, H. (1977) *But Now We Want the Land Back: A History of the Australian Aboriginal People*, Sydney: New Age Publishers.

Mitchell, I.S., and Cawte, J.E. (1977) 'The Aboriginal Family Voluntary Resettlement Scheme: an approach to Aboriginal adaptation,' *Australian and New Zealand Journal of Psychiatry*, 11(1): 29–35.

Morgan, G. (1999) 'The moral surveillance of Aboriginal applicants for public housing in New South Wales,' *Australian Aboriginal Studies*, 1999(2): 3–15.

New South Wales Parliament Legislative Council (1849) *Report of the Select Committee on the Aborigines and the Protectorate*, Sydney: Government Printer.

Reynolds, H. (1994) 'Origins and implications of Mabo: an historical perspective,' in W. Sanders (ed.) *Mabo and Native Title: Origins and Institutional Implications*, Canberra: Centre for Aboriginal Economic Policy Research, The Australian National University.

Rowley, C.D. (1970) *The Destruction of Aboriginal Society*, Ringwood: Penguin Books.

Sanders, W. (1984) 'Aboriginal town camping, institutional practices and local politics,' in J. Halligan and C. Paris (eds) *Australian Urban Politics*, Melbourne: Longman.

Sanders, W. (1995) 'Local governments and Indigenous Australians: developments and dilemmas in contrasting circumstances', *CAEPR Discussion Paper No. 84*, Canberra: Centre for Aboriginal Economic Policy Research, The Australian National University.

Smith, L.R. (1980a) 'New Black Town or Black New Town: the urbanization of Aborigines,' in I.H. Burnley, R.J. Pryor and D.T. Rowland (eds) *Mobility and Community Change in Australia*, St. Lucia, Queensland: University of Queensland Press.

Smith, L.R. (1980b) *The Aboriginal Population of Australia*, Canberra: Australian National University Press.

Taylor, J. (1997) 'Changing numbers, changing needs? A preliminary assessment of Indigenous population growth 1991–96', *CAEPR Discussion Paper No. 143*, Canberra: Centre for Aboriginal Economic Policy Research, The Australian National University.

Taylor, J. (2001) 'Aboriginal urbanisation,' in J. Jupp (ed.) *The Australian People: An Encyclopedia of the Nation, Its People and Their Origins*, Cambridge: Cambridge University Press.

Taylor, J. and Bell, M. (1996) 'Mobility among Indigenous Australians,' in P.W. Newton and M. Bell (eds) *Population Shift: Mobility and Change in Australia*, Canberra: Australian Government Publishing Service

Taylor, J. and Bell, M. (1999) 'Changing Places: Indigenous population movement in the 1990s', *CAEPR Discussion Paper No. 189*, Canberra: Centre for Aboriginal Economic Policy Research, The Australian National University.

11 Myth of the "walkabout"

Movement in the Aboriginal domain[1]

Nicolas Peterson

One of the much mythologized aspects of Australian Aboriginal behaviour has been the "walkabout", usually understood in terms of some internal urge that results in Aboriginal people leaving a locality without notice to travel for travel's sake.[2] It was a concept often called on by pastoral station managers, and other employers of Aboriginal labour across the continent, to account for the disappearance of members of their workforce without notice, or their unexpected reappearance. The mystery surrounding such movements was largely of the employers' own making: their overweening self-interest and complete lack of concern for Aboriginal people as social persons meant that few Aboriginal people could expect a request for leave to attend a ceremony, or to visit kin, to be granted. Indeed, until the middle of the twentieth century the police in many areas of north Australia, were used to track down and return Aboriginal workers who left a cattle station outside the wet season lay-off period.

Walkabout thus has elements of an everyday form of resistance: Aboriginal people avoided letting employers know when they intended leaving because they denied the employers' right to control their lives. Further, the employers' assumption that the urge to leave was biologically based helped reproduce unheralded departures because such an assumption meant that Aboriginal people were rarely called on to account for their movements. Since the need to travel was not biologically based it is of interest to explore some of the reasons for movement.

Pre-colonially, of course, varying degrees of movement were required in the pursuit of survival.[3] The nature and scope of this movement varied widely from coastal populations and those living on rich wetlands, who engaged in highly localized movement, to those in the harshest desert areas where movement was frequent and often over considerable distances. To this extent, pre-colonial population densities can be taken as an indirect mobility index. In the richly resourced areas, such as the estuarine environments of northern Australia, they could be as high as one person to one square kilometre, while in the most barren areas of the central desert, densities could drop to one person per 200 square kilometres. But it is clear that people also moved for a wide range of other reasons, sometimes over great distances (Young and Doohan 1989). The largest distances seem to have been in connection with certain kinds of trading (Mulvaney 1976), but other movements were connected with attending ceremonies, avoiding conflict or simply visiting relatives.[4]

Mobility is fundamental to an Aboriginal individual's social identity. Myers (1986) has traced the way in which Aboriginal social life in central Australia is underwritten by a tension between relatedness and autonomy. People have to work at producing relatedness and shared identity through visiting, participating in ceremonial life, marriage arrangements and exchange, constantly renewing their networks. Yet at the same time adult men and women place great emphasis on their personal autonomy, rejecting attempts by others to control or direct them, often solving the problem by moving. Although he was writing of people in the central desert his analysis applies across the continent, and is as true today of people living in the metropolitan and rural centres of the more settled areas of Australia as it is of people living in remote and sparsely settled parts of the continent.

Paradoxically, perhaps, for a people who were traditionally hunter-gatherers and moved around a lot, relationships to place and country were central to Aboriginal personal identity. In many areas of the continent the home area was believed to be the repository of spirit children from whom members believed they grew, and the landscape features to be the product of the activities of their heroic ancestors. While movement was intrinsic to the hunting and gathering way of life it was linked with a high value given to returning to the home area and, in particular, in the environmentally richer regions, to staying put. Following colonization many groups had to take up residence, scores if not hundreds of kilometres, away from their home countries without access to adequate transport, thus threatening their intimate psychological and life-giving links with them. It is illustrative of the intense emotional ties to country that in one desert area the Aboriginal people who only moved out of their country at adulthood, claimed to keep contact with it by making night time journeys to their home territories in dream-spirit form, flying through the air astride skeins of hair-string or on sacred boards (Tonkinson 1970).

In this chapter I will look at contemporary patterns of mobility that relate to Aboriginal domains in both remote and settled Australia.[5] I will begin with a discussion of the notion of domain and then examine a range of movements specifically related to it.

Aboriginal domains

John von Sturmer (1984: 219) has observed that:

> [i]n parts of remote Australia it is possible to talk of Aboriginal domains, areas in which the dominant social life and culture are Aboriginal, where the major language or languages are Aboriginal,

where the dominant religion and world views are Aboriginal, where the system of knowledge is Aboriginal; in short, where the resident Aboriginal population constitutes the public.

This concept has been taken up and elaborated on by David Trigger who defines such domains in terms of distinctive spheres of thought, attitudes, social relations and styles of behaviour (Trigger 1986: 99; see also Rowse 1992).

Although these domains usually have a spatial expression, they only equate with physical space where the spaces are clearly the province of Aboriginal social processes. In geographically remote parts of Australia the former Aboriginal reserved lands, which are now Aboriginal lands, have greatly facilitated the maintenance and reproduction of largely separate arenas or domains of social life. To a lesser extent this is also true of the more closely settled areas of Australia where Aboriginal people often live in separate villages built on former Aboriginal reserves on the fringe of country towns or in enclaves within towns or metropolitan centres. Trigger (1992: 100–2) argues that the issues at play find their context in unfolding power relations and the limiting of the intrusion and influences of non-Aboriginal people and institutions in Aboriginal lives.

Although there is an element of resistance in the maintenance of these domains by Aboriginal people, the separation was initially imposed on them by state legislation, which was in force until the late 1950s when it started to be modified. Land rights legislation, and now native title rights, have ensured the reproduction of many of these separate communities under Aboriginal control.

Prior to sedentarization the basic residential unit was a group of households that made up a band which occupied a range. The band was integrated into a regional network through the personal, social, political and ceremonial ties of individuals to other individuals in nearby bands. Aboriginal people speak of this network in a spatial way as being made up of "one-countrymen" which can be understood as a term, delineating the widely extended set of persons with whom one might reside and co-operate (Myers 1982: 180).

The loose integration at the regional level of one-countrymen/community was maintained, prior to settling down, by everyday social processes of visiting, trading, fighting, marrying and the less frequent but more significant activity of holding any of a wide range of ceremonies. This regional ceremonial life is, and has been, complex and dynamic for a long time. In the period immediately prior to European arrival, and subsequently, a major dynamic has been the competition between prominent men to create their own social domains, usually within more or less common "one-countrymen" communities, by promoting ceremonial gatherings of which they are the focus. They do this in a number of ways but it is clear that acquiring new ceremonies by exchange (Kolig 1981; Petri

1979) is most important. In the 1940s, an Aboriginal man from the Kimberley region of northwestern Australia is reported as saying of a newly introduced cult: "Some people like goranara [Kurangara] to make themselves great. They think they can rule the others" (quoted in Widlok 1992: 123).

Although since the 1970s there has been no legislative constraint on Aboriginal population movement, the continuing importance of kinship to Aboriginal people across the continent combined with racism, and Aboriginal preference, have meant that Aboriginal people tend to live, work and certainly socialize, largely with other Aboriginal people.[6]

Beats, runs and lines

The first account to draw attention to the way that Aboriginal movement in closely settled parts of Australia took place within Aboriginal domains was conducted in western New South Wales (Beckett 1988). Writing about Aboriginal life in the late 1950s, Beckett noted that in the absence of any wider co-ordinating principles in Aboriginal society, face-to-face relations are of primary importance among them and that local communities were only saved from isolation by the direct contacts their members made by moving about the countryside (Beckett 1988: 119). When travelling, proximity was not the issue. People would prefer to go two hundred miles to where they had kin, close affines and were known rather than ten miles to a place where they were not known. According to Beckett (1988: 131) (for localities see inset in Figure 11.1):

> All Aboriginal people have "beats," areas which are defined by the situation of kin who will give them hospitality, within which they can travel as much or as little as they please, and where they are most likely to find spouses, Proximity is only a minor factor. When first working in Murrin Bridge, I was impressed by the fact that most of the people knew far more about Wilcannia – 200 miles away – and visited it more frequently, than they did Euabalong – only ten miles away – or Condobolin – only thirty. One explanation is that the Murrin Bridge people have lived near these last two places only since 1948, whereas their contacts with the Darling River people go back to 1934 and before; however, one might expect links to have developed after a decade. In fact, over the last seven years only one marriage has been contracted with Condobolin and only one with Euabalong, as against six with Wilcannia.

Kin and friends were scattered largely as a result of moves forced on some people, particularly those out of work, by government; these people then became separated from friends and relatives who had managed to find work locally (Beckett 1988: 122–3). Subsequently much of the movement

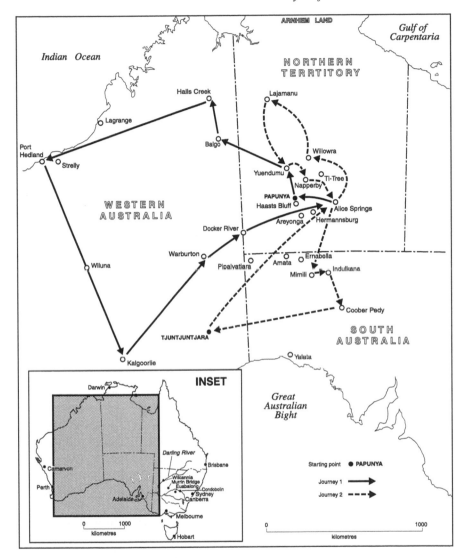

Figure 11.1 Jilkaja journeys in central Australia.

was by older women with grown-up children who spent their time visiting them in turn (Beckett 1988: 127). Because people from the one area tended to marry into the same small range of communities, the beats of many people from any one community coincided.

Similar patterns of movement have been reported from the southwest of Western Australia in the early 1980s (Birdsall 1988). As a result of World War II, the Department of Native Affairs was reduced in staff and unable to effectively administer and police the restrictive provisions of the 1936

Aboriginal Administration Act, which meant that Aboriginal people, known as Nyungars in this region, were able to move much more freely about the state in search of employment (Birdsall 1988: 137–8). Nyungars are organized in terms of cognatic family groups, which are spoken of as "all-one-family," and generally number between 200 and 300 individuals. The all-one-families are divided into a number of sections of approximately 40–150 people and are usually descended from a set of female sibling-cousins of late middle age. It is these women who work to hold their sub-division of the all-one-family together and it is within this group that members find their closest friendships and where their strongest loyalties lie. The members are likely to be scattered across a number of towns in the southwest, usually in one region with which the family has had a long history of association, and each family group refers to the set of towns where their members live as their "run" (Birdsall 1988: 141).

Other Nyungar families' longstanding connections with particular towns or regions have been disrupted in the past by forced movement to missions or government settlements far distant from their places of birth. In many cases these people have attempted to return to their old homes many years later, gradually working their way back moving from one town to another leaving some members along the way as they marry locally or take up permanent residence for one reason or another. This process has created "lines", some of which find members of an all-one-family group strung out over 2,000 km in towns along the Western Australian coast and hinterland. Unlike the run, these lines do not constitute an ancestral claim to an entire region (Birdsall 1988: 142).

These lines and runs reflect the centrality of membership of a family group to Nyungar social identity. This membership is based not just on consanguinity but on having been reared by one of the senior women of the family. Affines can be linked to an all-one-family group through somebody who has been reared in this way (Birdsall 1988: 143). Rearing relationships are established through the practice of sending or taking children to live with their mother's kin for varying periods of time throughout their childhood which leads to children coming to form individual preference for particular matrilateral kin. An individual who fails to regularly visit and be visited by kin both within their town or along their run or line will eventually lose their close relationship to their sibling-cousins and be deprived of wide social support (Birdsall 1988: 145).

Young men can travel alone or in groups, but young women rarely ever do so. Usually women travel in groups that span three generations and cars are much preferred over the bus. Birdsall (1988: 146–7) provides details on the travel frequency of members of one family group of 162 people whose line runs from the metropolitan area of Perth to the regional centre of Carnarvon 800 km to the north. Over an eight-week period in June–July 1982, 38 people engaged in 24 separate trips travelling to, from, or through one of the towns on the line. That is to say that in this period more than 23 per cent of the family members were known to be in transit somewhere along the line.[7] As Birdsall (1988: 156) says, "It is through the practice of regular, frequent travel that the unity of the family community and the continuity of Nyungar social organization are maintained over the distances of time and space which separate the members of the community from one another".

Other accounts of coming and going

As Annette Hamilton (1987: 47) has observed, little attention has been paid to understanding the nature of and reasons for patterns of Aboriginal movement, yet its frequency when documented is often striking. In a small northern South Australian community of around 40–60 people, in which she worked during 1970–1971, there was a change in the numbers of people on 37 of 62 consecutive days in the months of July and August with numbers fluctuating from a low of 13 to a high of 57 (Hamilton 1987). On 37 days there were visitors in camp but not all changes in the numbers were due to visitors as some were due to residents leaving to make visits elsewhere. Hamilton (1987: 49) noted that:

> Aboriginal people will, at no notice, join a vehicle travelling hundreds of kilometers away, taking with them no money and few provisions, and will have no idea of when or how they will return . . . [on two conditions] firstly that the vehicle is in the charge of a relative, and secondly that relatives can be found at the end-destination and intermediate stops along the way. Travelling from place to place can only be undertaken in this apparently haphazard way precisely because an elaborate network of reciprocal exchanges underpins it, where by relatives accept unannounced visits from one another and provide the wherewithal for the visitor's survival if necessary.

Hamilton (1987: 49) argues that this movement should not be seen as an example of cultural continuity based on a direct preservation of past mobility patterns, but as a means of ensuring social and economic survival in the absence of fully articulated mechanisms of social and economic incorporation into the developing frontier society.

Jon Altman (1987: 22–3) provides general evidence for high levels of mobility in an Arnhem Land outstation in the period between October 1979 and October 1980, providing a standard deviation of mean monthly populations which he calls the "mobility factor". For a mean monthly population that ranged between 18.3 and 43.6 the mobility factor ranged from 2.7 in the wet season to 18.7 at the beginning of the dry. This population was living in an entirely Aboriginal domain, the then Arnhem Land Aboriginal Reserve. He argues that mobility had been intensified at the time of his fieldwork because of the availability of vehicles, the existence of a regional service centre forty kilometres away that seemed to exert a pull effect, and the availability of social security incomes which people spent at the centre (Altman 1987: 104). Much of the movement that was not for subsistence purposes or visiting the centre was connected with attending ceremonies.

In a recent study of mobility in the Aboriginal domain on Cape York Peninsula, Ben Smith looks at what happens when a pre-colonially semi-nomadic population, that has long been sedentized, regenerates a shifting pattern of mobility based on small family groups with the establishment of small decentralized settlements known as outstations in the 1990s (Smith 2000: 13, 65). Through the use of quantitative and qualitative data he relates this mobility to the central importance of kinship in Aboriginal sociality and changing relations to land under the impact of the recognition of native title rights and state intervention.

Contemporary ceremonial movement

While movement to attend ceremonies remains common in remote Australia, one form of ceremonial movement in the Aboriginal domain is particularly significant. This is the central Australian initiation journey, which with access to vehicles since the late 1960s has increased in length and often in the numbers of people involved (Peterson 2000). According to customary practice, just prior to circumcision, boys were taken by their guardians on a journey to visit people in the region to gather them together for the actual ceremony. In the past this was done on foot over several months; today some initiation candidates, but not all, go off on a tour with their guardian using various forms of transport. While there is no specific evidence for the number of such journeys in any year it is probable, taking the vast central desert regions as a whole with its large number of small communities, that there would be half a dozen or so each year, although most journeys if not all would be substantially shorter than that described in the following case study.

Early in the spring of 1994 a Western Desert circumcision candidate set out on such a pre-initiation journey to gather people for his ceremony. What was remarkable about this particular journey was the distance covered: it took the boy and his guardian over 2,250 km on their outward journey. At that point they turned round to retrace their steps gathering people along the way in a party widely known as *jilkaja*. By the time the *jilkaja* party arrived back at the boy's home community over 600 people were travelling with the initiate.

Prior to the 1960s, Aboriginal people in remote Australia had very limited access to transport and would make these journeys on foot, by donkey, camel, horse or even bicycle (Petri 1979: 232), but from the middle of the 1960s they increasingly used the car (Nungarrayi 1995). The first motor vehicle owned by a Pitjantjatjara man on the eastern side of the desert was a second-hand Land-Rover purchased in 1961 with wages from employment on a water drilling rig (Edwards 1994: 148). On the western side of the desert at Warburton cars started to appear around 1966 with the opening up of a mine near the mission (Gould 1969: 171–2). By 1970 – a period that Hamilton refers to as a bridge between the use of camels and cars – cars were becoming more common in the Pitjantjatjara area (Hamilton 1987: 51; Peterson 1977: 144) and in the west around Wiluna (Sackett 1980–1982: 130).

The watershed was 1968 because in that year the Department of Social Security started paying allowances in cash directly to the beneficiaries rather than to the superintendent of the village in which they lived: it was this access that led to the instant abandonment of camels, used by some Pitjantjatjara groups up until that time, and a greatly increased involvement with cars.

Akerman (1979: 236–8) has noted that there was a marked increase in ritual activity in the Kimberley around the same time (1972–1978) and mentions access to motor vehicles as one important factor in this increase. Today, Aboriginal people generally have much greater access to vehicles but keeping them in running order remains a problem and most vehicles have quite short lives.

The recent history of *jilkaja* journeys is not easily documented. Even where large numbers of people descend on a village for a few days it is only when there is a logistical problem for the European staff that they appear to make any lasting impression. Otherwise such movements are common enough that they attract only passing attention from non-Aboriginal people working closely with Aboriginal people.

For the general public such movements largely go unseen since they mainly take place on back roads, firmly on Aboriginal lands and within Aboriginal domains, and where they leave these areas such large groups of Aboriginal people together are seen as intimidating and are avoided. For Aboriginal people in any community they have a marked influence on patterns of movement. Roads become closed to Aboriginal movement for nobody should meet a *jilkaja* party on the road or leave a community before it, especially women and uninitiated males. As a consequence roads may be devoid of any Aboriginal travellers for days at a time in anticipation of a party arriving.

The following information on specific *jilkaja* journeys is fragmentary but indicative. The first documented large gathering based on the car was at Docker River on the border of the southwest of the Northern Territory and Western Australia in October 1972. In that month circa 1,700 people descended on the village in 55 cars and seven trucks for a week (Peterson 1977: 146). Sackett (1978: 111) reports that Wiluna residents travelled 1,385 km to Docker River in that year, presumably to this ceremony. He also indicates that they travelled 1,000 km to Warburton in 1973 and 1,090 km to Strelly in 1974. Hope (1983: 80) shows the range of *jilkaja* journeys made from Amata for a twelve-year period from 1968, which included Yalata, Warburton, Papunya and Indulkana at the extremities. This evidence appears to confirm the early and strong involvement of Pitjantjatjara and people to the west with the expansion of the journeys through their early access to cars.

At Yuendumu, on the eastern fringe of the greater Western Desert area access to cars was slower to develop. At the time I commenced fieldwork in Yuendumu in 1972, Aboriginal people had been integrated into the cash economy for about four years. Most of this money came from social security and the rest from a government commitment to provide a job for any Aboriginal person who wanted one on a training allowance which ran at about one-third of the minimum wage. Thus, although people were not wealthy they started to be able to afford second-hand cars through accumulation of money in card games or through the pooling of funds by close kinsfolk. These vehicles were nearly always of poor quality and had a short useful life as the huge dead car dumps at the fringe of all villages attested.

There is limited information on the extent of *jilkaja* journeys prior to 1972 in the Yuendumu area, although Meggitt (1962: 285) does indicate that in the early 1950s initiates from Yuendumu visited Mt Doreen, Mt Denison and Coniston cattle stations, all close by, and occasionally Haasts Bluff to the south (see also Long 1970: 329). For most of 1972–1973 there were only two regularly working cars owned among one thousand Warlpiri people, although a number of others put in brief appearances of a week or two before breaking down irreparably. But with more money around, the Yuendumu community responded to the visit of an initiate from Papunya, 90 km to the south, in February 1973 by having a collection on pay day of $A7 per head to hire three buses. One-hundred-and-twenty-one men bought tickets and a few additional people went with me in my four-wheel drive. The *jilkaja* stayed three days before returning to Papunya accompanied by the above party. Later on that year a relatively small group hired a lorry that took them to LaGrange, 1,200 km to the northwest, where they stayed two nights before coming back. It was the first time that many on the journey had seen the sea.

This expansion of the catchment of initiation ceremonies was not just dependent on access to vehicles but also a product of expanded, government inspired meetings on policy issues and, more importantly, sporting events that followed the election of the federal Labor government in 1972 and a radical shift in policies. In particular the Yuendumu Sports Weekend, a central Australia-wide sports carnival held each August, grew rapidly through the 1970s. In 1973 Pitjantjatjara men and Warlpiri men who were total strangers met for the first time at the Yuendumu Sports Weekend. The next year initiation ceremonies at Yuendumu had over 700 men present, some of them Pitjantjatjara men from Docker River. In the Pitjantjatjara area the major sports carnival held at Amata in July 1976 was a turning point in the unification of the Pitjantjatjara who formed the Pitjantjatjara Council immediately following that event (Edwards 1983: 297). Improvements in the outback road system and the growth in the availability of radio and telephone services to Aboriginal people have also played a significant part in the organization and growth of the Aboriginal domain (Stanton 1983: 171).

In April 1976 a major initiation ceremony was held in Papunya with people in attendance from the communities of Yuendumu, Hermannsburg, Areyonga, Indulkana, Ernabella, Mimili, Amata, Pipalyatjara and Docker River. The youth only travelled to Indulkana: from there on the hairstring belt worn by the novice was sent summoning people. Over 600 men were involved in the ceremony and three youths were circumcised.

The following month another novice was sent out from Papunya, this time on a northerly circuit to Yuendumu, Willowra, Balgo and probably Lajamanu. The differences between these two circuits suggest that the April 1976 novice was a Pintupi or Pitjantjatjara youth while the May 1976 youth was probably Warlpiri.

In 1977, or 1978, a senior Pintupi man set out with a novice and a driver from Papunya with enough food and petrol to get to Yuendumu (see Journey 1 in Figure 11.1). From Yuendumu he travelled to Balgo, Halls Creek, Port Hedland, Kalgoorlie, Warburton, Docker River, Alice Springs and back to Papunya. This route was unusual in that being a circuit the novice would not have brought everybody attending back with him. But it is typical in the way that one aspect of these journeys is funded: the novice and guardian(s) are supported by each community they turn up to and it is the senior members of that community that pass them on to senior members of the next community.

Even from this limited information it is clear that with access to the car and the new networks created partly as a result of the increased catchment of meetings and sporting events, that the *jilkaja* journeys expanded rapidly from the beginning of the 1970s and the numbers of people attending some initiations also grew substantially.

Jilkaja *journey to Tjuntjuntjara*

An initiation journey made in October 1994 started at Tjuntjuntjara outstation in Western Australia with the novice and guardian flying to Alice Springs and then on to Willowra (see Journey 2 in Figure 11.1). A leading man at Willowra then handed the novice and his guardian on to the Lajamanu community: by road Lajamanu is approximately 2,250 km from Tjuntjuntjara. At Lajamanu the novice started the journey back. He was accompanied by two buses, one small car, one Toyota Landcruiser and a large truck filled with supplies. At Yuendumu twelve cars joined the party.

By the time they reached Alice Springs they had been joined by people from Napperby station, Willowra and the nearby Aboriginal owned station

of Ti-Tree and people from the Papunya area. They camped in Alice
Springs for two days securing money from royalty accounts set up to assist
with ceremonial expenses.

They then drove south to Mimili and Indulkana staying two days in the
latter place. From Indulkana they went to Coober Pedy, then drove south
towards the railway line and turned west along a back country road until
they arrived at Tjuntjuntjara.

By the time the convoy reached Tjuntjuntjara it had more than 30 vehicles
(cars, buses and trucks) carrying some 300 men and a similar number of
women and children. They joined over 400 people gathered at Tjuntjuntjara,
the normal population of which is around 200, with the eventual estimated
total number of people present around 1,200 people. The Northern Territory group stayed there three or four nights during which time a number of
young men were circumcised.

Through the extension of the initiation journey, the greater Western
Desert area is gradually being integrated into a common moral community,
based on pre-existing broad commonalities of language and culture and
improved access to transport. In the widespread Pitjantjatjara form of the
ceremony, which dominates much of the greater Western Desert, the age-
set-like character cuts across locality and gives prominence to younger men
in their thirties as the organizers and key participants. These ceremonies
are open-ended and inclusive, based on knowledge and understandings
that are widely held by all adult men.

The conjunction of increased independent access to vehicles, with the
ever-present desire to expand relatedness, has meant the initiation ceremonies are the ideal means for extending social relations in an egalitarian
way. It is quite clear that all men get great pleasure from the travelling and
more importantly from the ceremonies themselves.

Despite the greatly increased mobility most Aboriginal people still remain
firmly embedded in their home village, even if there is high mobility
between the villages and the regional town centres. Kinship, cultural practice,
geography and racial difference ensure the continued reproduction of the
village communities.

Conclusion

In 1994, the Australian Bureau of Statistics (ABS) carried out a national
survey of Aboriginal and Torres Strait Islander people. A limited number of
questions relating to "culture" were asked, including some relating to
participation in indigenous cultural activities. Cultural activities were defined
as: ceremonies, funerals, festivals and carnivals and those involved with
Aboriginal organizations. It is assumed here that the first two activities were
entirely Aboriginal domain activities whereas the other two, while within the
Aboriginal domain are, in effect, largely state sponsored activities in one way
or another. Fifty-four per cent of respondents, aged 13 years and over, had

attended a funeral in the previous year. The rates varied between capital cities where the percentage was 36 per cent, other urban areas 58 per cent, and rural Australia where it was 66 per cent. Attendance at ceremonies for the same three groupings was 10 per cent, 17 per cent and 35 per cent respectively (ABS 1994: 8). Not all of these cultural activities would have required all participants to have made a journey over some distance but many would have done. Indeed, only 2 per cent, 4 per cent and 7 per cent respectively of people in these three categories said they could not attend a cultural activity because it was too far away and 4 per cent, 4 per cent and 9 per cent respectively said this was because of lack of transport (ABS 1994: 8).

In respect of attendance at funerals the participation of males and females was identical at 54 per cent but in the case of ceremonies a slight difference emerged with 22 per cent of males but only 20 per cent of females attending. In respect to obstacles to attending cultural activities women were slightly more disadvantaged than males by lack of transport and distance: 6 per cent compared to 5 per cent, and 5 per cent to 4 per cent respectively.

Writing on the nature of native Hawaiian mobility, Jonathon Friedman describes it in terms of endosociality by which he is referring to the way, when travelling, they manage to avoid contact with the larger society, sitting with their backs to the world "as if in conduits of their own making. Home is thus extended to the larger region via a complex of insulated networks" (Friedmann 1997: 284). Although it is now illegal for Aboriginal people to travel sitting in the back of open trucks in remote Australia, much Aboriginal travel does, metaphorically, take place with their backs to the world, turned in on their own domains. This is true of Aboriginal people throughout the continent, although to different degrees. Aboriginal endosociality is partly a product of the same racism that gave rise to the myth of walkabout but it is also a product of a distinctive culture with its egalitarian ideology, its emphasis on the relational constitution of the person and the importance of place in the constitution of personal identity. Together these factors continue to underwrite frequent mobility in the Aboriginal domain.

Notes

1 I have received assistance and information from a large number of people in preparing this chapter. I would particularly like to thank H. Nelson, B.L. Japaljarri, H.C. Jakamara, and D. Japangardi for telling me about the journey to Tjuntjuntjara. For an historical perspective on initiation journeys by car I am much indebted to Dick Kimber's wonderfully careful and detailed correspondence from Central Australia over a quarter of a century. I would also like to thank Scott Cane, Bill Edwards, Derek Elias, Philip Jones, David Nash, Lee Sackett, John Stanton, Bob Tonkinson, Graham Townley, David Trigger, Peter Twigg, and Thomas Widlok for various kinds of valued comment and assistance.
2 Arthur (1996: 174) gives two meanings of "walkabout". The first is: "originally was applied to travel, usually in the traditional manner . . . most commonly

been applied to the lay-off season in the northern cattle industry, when Aboriginal employees (most often living in their own country) would leave the station for a period of traditional life". The second meaning is when it is "used of, rather than by, Aboriginal people it usually implies an abandoning of responsibilities, rather than a return to Aboriginal responsibilities". An example of this usage is given in Sorenson (1911: 43): "She leaves the station with her followers pretty frequently for a 'walk-about,' for the call of the wild comes irresistibly, no matter how long she has mixed with the whites". While this quotation refers to a woman it is probably true that it is the men that were and are the more mobile. The Australian Geographical Society, an educational and scientific organization formed with the object of promoting geographical knowledge of Australia and adjoining islands', published a well known and widely circulated monthly journal called *Walkabout Magazine* and provided an explanation of the magazine's title beneath the mast head: "The title, 'Walkabout', has an 'age-old' background and signifies a racial characteristic of the Australian aboriginal who is always on the move. And so, month by month, though the medium of pen and picture, this journal will take you on a great 'walkabout' throughout Australia and the islands of the South-west Pacific". *Walkabout Magazine*, 20(5): 1 May, 1954: 9.

3 For a classic early description of seasonal movement by a single group and an account of how the remains left behind could lead to the belief that people from two different cultural groups were involved see Thomson (1939).

4 C.W. Hart provides an account of the annual activities of an active Tiwi elder at a time in 1928–1929 when 75 per cent of the Tiwi population led an independent self-supporting life in the bush. This man spent 39 weeks alone with his own extended family group. In the other thirteen weeks he attended three big funerals, one initiation ceremony, several small funerals and made a few other visits to neighbouring districts. He also hosted a joint kangaroo hunt in his own area, attended the naming ceremony of one of his daughters' children and was a member of one war party (see Hart *et al.* 1988: 47–9).

5 By settled Australia I refer to the area east of a line from Cairns to Port Augusta and the southwestern portion of Western Australia, west of a line from Carnarvon to Esperance. The remaining part of Australia is the area referred to as remote. Rowley (1971) provides a map showing the two regions. I do not deal with the micro-movements within camps or villages which often take place on a daily basis (see Martin and Taylor 1996; Musharbash 2000: 59) or result from vacating a house on death of an occupant in remote Australia.

6 Although the 1996 census shows that 64 per cent of Aboriginal adults in couple relationships have a non-Aboriginal partner (Taylor 1997). This high level of out-partnering is relatively recent and there is some reason to believe that the non-Aboriginal spouses get drawn into Aboriginal social networks to a greater extent than the other way round.

7 Some other members of the family were known to be travelling but not through the town where Birdsall was based and she was unable to accurately determine the numbers (Birdsall 1988: 147).

References

Akerman, K. (1979) 'The renascence of Aboriginal law in the Kimberleys', in R.M. and C.H. Berndt (eds) *Aborigines of the West: Their Past and Their Present*, Perth: University of Western Australia Press.

Altman, J. (1987) *Hunter-Gatherers Today: An Aboriginal Economy in North Australia*, Canberra: Australian Institute of Aboriginal Studies.

Arthur, J.M. (1996) *Aboriginal English: A Cultural Study*, Melbourne: Oxford University Press.

Australian Bureau of Statistics (ABS) (1994) *National Aboriginal and Torres Strait Islander Survey 1994: Detailed Findings*, Canberra: Australian Bureau of Statistics.

Beckett, J. (1988) 'Kinship, mobility and community in rural New South Wales', in I. Keen (ed.) *Being Black: Aboriginal Cultures in 'Settled' Australia*, Canberra: Aboriginal Studies Press.

Birdsall, C. (1988) 'All one family', in I. Keen (ed.) *Being Black: Aboriginal Cultures in 'Settled' Australia*, Canberra: Aboriginal Studies Press.

Edwards, B. (1983) 'Pitjantjatjara landrights', in N. Peterson and M. Langton (eds) *Aborigines, Land and Land Rights*, Canberra: Australian Institute of Aboriginal Studies.

Edwards, B. (1994) 'Mutuka Nyakunytja – seeing a motor car', *Aboriginal History*, 18(2): 145–58.

Friedman, J. (1997) 'Simplifying complexity: assimilating the global in a small paradise' in K. Olwig and K. Hastrup (eds) *Siting Culture: The Shifting Anthropological Object*, London: Routledge.

Gould, R. (1969) *Yiwara: Foragers of the Australian Desert*, London: Collins.

Hamilton, A. (1987) 'Coming and going: Aboriginal mobility in north-west South Australia, 1970–1971', *Records of the South Australian Museum*, 20: 47–57.

Hart, C., Pilling, A. and Goodale, J. (1988) *The Tiwi of North Australia* (3rd edition), New York: Holt, Rhinehart and Winston.

Hope, D. (1983) 'Dreams contested: a political account of relations between South Australia's Pitjantjatjara and the Government, 1961–1981', unpublished PhD thesis, Adelaide: Flinders University.

Kolig, E. (1981) *The Silent Revolution: the Effects of Modernization on Australian Aboriginal Religion*, Philadelphia: ISHI.

Long, J. (1970) 'Change in an Aboriginal community in central Australia', in A. Pilling and R. Waterman (eds) *Diprotodon to Detribalization: Studies of Change among Australian Aborigines*, East Lansing: Michigan State University Press.

Martin, D. and Taylor, J. (1996) 'Ethnographic perspectives on the enumeration of Aboriginal people in remote Australia', *Journal of the Australian Population Association*, 13(1): 17–31.

Meggitt, M. (1962) *Desert People*, Sydney: Angus and Robertson.

Mulvaney, D. (1976) 'The chain of connection', in N. Peterson (ed.) *Tribes and Boundaries in Australia*, Canberra: Australian Institute of Aboriginal Studies.

Musharbash, Y. (2000) 'The Yuendumu community case study', in D.E. Smith (ed.) *Indigenous Families and the Welfare System: Two Community Case Studies*, Canberra: Centre for Aboriginal Economic Policy Research, The Australian National University.

Myers, F. (1982) 'Always ask: resource use and land ownership among Pintupi Aborigines of the Australian Western Desert', in N. Williams and E. Hunn (eds) *Resource Managers: North American and Australian Hunter-Gatherers*, Boulder, CO: Westview Press.

Myers, F. (1986) *Pintupi Country, Pintupi Self: Sentiment, Place, and Politics among Western Desert Aborigines*, Washington DC: Smithsonian Institution Press.

Nungarrayi, M. (1995) 'On travelling through country: from foot to motorcar', in P. Vaarzon-Morel (ed.) *Warlpiri Women's Voices*, Alice Springs: Institute for Aboriginal Development Press.

238 *Nicolas Peterson*

Peterson, N. (1977) 'Aboriginal involvement with the Australian economy in the Central Reserve during the winter of 1970', in R.M. Berndt (ed.) *Aborigines and Change: Australia in the '70s*, Canberra: Australian Institute of Aboriginal Studies.

Peterson, N. (2000) 'An expanding Aboriginal domain: mobility and the initiation journey', *Oceania*, 70(3): 205–18.

Petri, H. (1979) 'Pre-initiation stages among Aboriginal groups of north-west Australia', in R.M. and C.H. Berndt (eds) *Aborigines of the West: Their Past and Their Present*, Perth: University of Western Australia Press.

Rowley, C.D. 1971 *The Remote Aborigines*, Canberra: Australian National University Press.

Rowse, T. (1992) *Remote Possibilities: The Aboriginal Domain and The Administrative Imagination*, Darwin: North Australia Research Unit, The Australian National University.

Sackett, L. (1978) 'Punishment in ritual: "man-making" among Western Desert Aborigines', *Oceania*, 49(2): 110–27.

Sackett, L. (1980–1982) 'Working for the law: aspects of economics in a Western Desert community' *Anthropological Forum*, 5(1): 122–32.

Smith, B. (2000) 'Between places: Aboriginal decentralisation, mobility and territoriality in the region of Coen, Cape York Peninsula (Queensland, Australia)', unpublished PhD thesis, London: London School of Economics, University of London.

Sorenson, E.S. (1911) *Life in the Australian Backblocks*, Melbourne: Whitcombe and Tombs.

Stanton, J. (1983) 'Old business, new owners: succession and "the Law" on the fringe of the Western Desert', in N. Peterson and M. Langton (eds) *Aborigines, Land and Land Rights*, Canberra: Australian Institute of Aboriginal Studies.

Taylor, J. (1997) 'Policy implications of indigenous population change, 1991–1996', *People and Place*, 5(4): 1–10.

Thomson, D. (1939) 'The seasonal factor in human culture: illustrated from the life of a contemporary nomadic group', *Proceedings of the Prehistoric Society*, 5: 209–21.

Tonkinson, R. (1970) 'Aboriginal dream-spirit beliefs in a contact situation: Jigalong, Western Australia', in R.M. Berndt (ed.) *Australian Aboriginal Anthropology*, Perth: University Western Australia Press.

Trigger, D. (1986) 'Blackfellas and whitefellas: the concepts of domain and social closure in the analysis of race relations', *Mankind*, 16(2): 99–117.

Trigger, D. (1992) *Whitefella Comin': Aboriginal Responses to Colonialism in Northern Australia*, Cambridge: Cambridge University Press.

von Sturmer, J. (1984) 'The different domains', in *Aborigines and Uranium: Consolidated Report on the Social Impact of Uranium Mining on the Aborigines of the Northern Territory*, Canberra: Australian Institute of Aboriginal Studies.

Widlok, T. (1992) 'Practice, politics and ideology of the "travelling business" in Aboriginal religion', *Oceania*, 63(2): 114–36.

Young, E. and Doohan, K. (1989) *Mobility for Survival: A Process Analysis of Aboriginal Population Movement in Central Australia*, Darwin: North Australia Research Unit, The Australian National University.

12 The social underpinnings of an "outstation movement" in Cape York Peninsula, Australia

Benjamin Richard Smith

Despite the substantial impact of colonialism on the Indigenous population of central Cape York Peninsula and resulting sociocultural transformation, population mobility remains at the heart of Aboriginal life across the region. This chapter seeks to explore the contemporary rates, forms and motivations of Aboriginal mobility among the Indigenous population associated with Coen, a township in central Cape York Peninsula, Australia.[1] Over the past decade, the region has seen a return to a high rate of population mobility after two decades of spatial concentration and sedentarization, this mobility being closely tied to the development of regional decentralization and the emergence of what has been generally labelled an "outstation movement". In common with a number of other areas across northern and central Australia over the past thirty years (see Chapter 2 and Davis and Arthur 1998), local Aboriginal people have established a number of small settlements, generally called "outstations", in a 100 km radius of the township of Coen (Figure 12.1) and the majority of the population now move frequently between these, the township of Coen and other regional population centres.

A review of data on "classical" mobility practices and recent qualitative and quantitative data drawn from the region and elsewhere suggests that this renaissance in Aboriginal population mobility has its foundations in a complex interplay between persistent aspects of cultural production, the local importance of originally exogenous practices – particularly those associated with the pastoral milieu – and the deepening involvement of the state in local Indigenous lives. Population mobility has allowed the mediation of these and other aspects of contemporary Aboriginal life and points to the attempts of local Indigenous people to shape both their day-to-day lives and broader emergent forms of post-colonial social life, and the conjoint attempts of local and wider administrative bodies to engage with these shorter- and longer-term processes. The resulting pattern of decentralization, in this region as elsewhere across the continent, marks a significant reversal of the wider trend towards urbanization in population movements by Indigenous peoples.

Figure 12.1 Coen and its outstations in Cape York Peninsula.

Historical impacts on mobility patterns

Early demographic data which might point to the forms of mobility "at the
threshold of colonisation" (Keen 2001) in the region are scarce, but data
from anthropological accounts of the region from the 1920s and 1930s, in
combination with later accounts, oral history and information from
elsewhere across the continent allow for the reconstruction of the kinds of
mobility likely to have occurred before substantial European impact and
the reasons for these patterns.

As has become increasingly apparent across the continent, the Aboriginal people of the region were not simply "nomadic", a persistent stereotype that has been collapsed even for the more arid areas of Australia.[2] Rather, as with other peoples whose economic and social lives are characterized by a relatively high rate of economic and residential mobility, Aboriginal Australian mobility in this region has remained based in a set of connections, including proprietary ownership (see Rigsby 1998), with tracts of land and particular places and resources located within them, and seasonal shifts between smaller more mobile bands and larger more sedentary camps. For these reasons, it may be better to characterize the economic lifestyle and its associated mobility practices as one of "shifting hunter-gatherers" rather than "nomads".[3]

Although it is difficult to reconstruct pre-contact Aboriginal society with any certainty, and more so for the inland regions of the Peninsula where information is sparser than for the coastal areas, it is apparent that despite being marked by a mobile, band-oriented lifestyle, the relation of Aboriginal people to place was far from the chaotic wandering perceived by early colonial writers. Donald Thomson, who worked on the coast 60 km east of Coen, recalled that:

> During the whole of my experience of the Yintjingga [a "tribe" camping in the locality of the same name], in 1928, and again in the following year, the horde at the Stewart River remained close to the estuary. From time to time small parties, generally consisting of a number of men or of one or two families, went inland for a few days, but the headquarters of the group remained at Yintjingga (Stewart River Estuary) . . . From time to time the camp moves bodily a short distance up or down the river – never as a body moving far from the mouth. In the period of approximately five months during which I was in close touch with the horde in 1928, six such moves were made – each of about two or three hundred yards.
>
> (Thomson 1934: 241)

Thomson does not mention that the group's mobility may have been affected by the impact of colonialism – the area was, at the time of his research, the site of the wharf servicing the Coen township, although Rigsby (1999: 109) notes that, by the time of World War I, the white settlement there had dwindled to a single wharf-keeper and Aboriginal people with ties to the area had returned and were camping "nearby along the beach esplanade and sandspits at the river mouth". However, work by Chase in the 1970s with groups who formerly ranged along neighbouring areas of the east coast indicates similar movement patterns in "pre-contact" coastal Aboriginal society (Chase 1980: 158–9). Chase describes the seasonal oscillation between smaller and larger camps along the coast. During the dry season, the lives of coastal groups were based around a number of "big camps", consisting of twenty or more campsites along 210 km of coast, of which around half would be in use at any one time during the dry season. Chase ("perhaps conservatively") estimates the population of each of these camps as around eighty people. The composition of these groups drew from a number of coastal (land-owning) "clan" or "country" groups,[4] as well as kin from inland "estates". The camps at these sites were moved, a few hundred metres at a time over an area of a few kilometres as resources permitted to compensate for fouling and the general pollution of camp sites (Chase and Sutton 1998/1981: 70–1).

Chase's informants described these camps as "'main' camps where people normally resided, but from which they could strike out, either as individuals or small family groups, on trips of short duration, to exploit particular resource concentrations on their own [patrilineal] estates" (Chase 1980: 157), and one of Chase's informants noted "might be *kayaman* [dry season] just family go walkabout, look for that yam, look at them story places, but he can't stay away. He come back and join up mob at main camp" (Chase 1980: 157).

Every few years there was a major congregation of the coastal groups for one of three sets of initiation ceremonies associated with the three main coastal languages. Each language-territory had one ceremonial ground where these rites were performed. The camps associated with these occasions were extremely large, and lasted for close to two months – the duration of the rituals performed there (Chase 1980: 162–3). These ceremonies were the basis for an unusual regional emphasis along the northeast coast, in comparison with other parts of the Peninsula (see Chase 1984).

In the wet season, these dry season camps were compressed into an even smaller number of locations: a total of twelve of these existed along the coastal strip which Chase mapped. These were places with dry terrain, access to basic resources "and a useable route or 'road' *(kayan)* either inland or to other beach areas" (Chase 1980: 158). In these larger camps, the less preferred (and more labour intensive) food which was eaten there "as well as the psychological strain of being camp-bound, crowded and living in near-constant rain" (Chase 1980: 156), produced heightened inter-personal tensions and frequent fights.

Although the forms of local organization in the inland part of the central Peninsula region are less certain, it is apparent that they were not dissimilar to those of the coast in many respects, save that they were markedly less regional, and more intensely localized in emphasis. In both regions there were spatially dispersed, patrilineally recruited land-owning groups or clans typically comprising fifteen to thirty people (Chase and Sutton 1998/1981: 69), and localized land-using bands. Similar to the coastal people, it is apparent that the lifestyles of these inland bands revolved around a number of focal campsites. The contemporary descendants of Ayapathu-speaking clans, for instance, note that Polappa, a place to the south of Coen noted for several large, perennial water sources was "the main camp place for Ayapathu people". As with the coastal population, the inland people's dry-season activities appeared to have been focussed around these "main camps", with small groups ranging across surrounding estates in order to utilize resources there, visit areas with which they had personal connections and perform "increase ceremonies" to ensure the proper management of their "country". People also moved between the larger camps, to attend initiation ceremonies, arrange marriages or visit kin. As one Ayapathu man put it:

> [c]attle roam like old Murri [Aboriginal people] before. Old people used to walk to Jabaroo, Rokeby, Archer [River], Aurukun, Buthen Buthen, Lockhart [River], Station Creek to Port Stewart, down to the coast. Walk to a different place for a change. Walk, rest up in different camp. Cattle the same way. First burn, walk from old grass to catch the new grass.
>
> (P. Port, personal communication, July 1996)

Colonial impacts on population mobility

With the impact of colonialism, the Aboriginal landscape and the movements and activities of Aboriginal people within it changed. Initial explorations, from 1848, were met with resistance although these initial intrusions effected little disruption in Aboriginal lives. But the discovery of gold on the Palmer River in 1872 rapidly flooded the southern Peninsula with thousands of European and Chinese prospectors. The prospectors worked their way north and in 1876 gold was discovered near the site of the future Coen township, leading to a brief but unsuccessful gold rush. The construction of a telegraph line through the centre of the Peninsula, the arrival of pastoralists from further south to supply the mining population and the re-establishment of the Coen goldfield saw the beginnings of real colonial impact in the region. The first cattle stations, founded in the 1880s, were all located at permanent waterholes that provided important camp-sites for Aboriginal people and the cattle-runs of the stations occupied surrounding country, spoiling water sources and disrupting the local ecosystem.[5] Fearful of Aboriginal attacks, the stations commonly drove Aboriginal people away from their properties and mounted revenge attacks for the spearing of cattle, supported by the murderous patrols of the Native Mounted Police (see Loos 1982; May 1994).

Simultaneously, mining camps and works sprang up across the inland centre of the Peninsula, again commonly sited near or at important resources and camping places. The impacts of malnutrition, disease and violence rapidly weakened Aboriginal resistance such that, by the early 1890s, local pastoralists began attempts to reach conciliation with local Aboriginal people, and in 1895 the subdued Aboriginal population around

Coen was considered safe enough to be "allowed in" to the Coen township. Aboriginal people sought to shift their lifestyles to incorporate the appearance of new resource centres, compensating for the loss of previous ones. Whites at telegraph stations, cattle stations, mining camps and the Coen township began to supply rations and approach a "conciliation" with the bedraggled Aboriginal population living around them. In some cases a monthly bullock-kill was instituted. Aboriginal people began to be incorporated within the white domain as a source of labour and near-permanent Aboriginal camps accompanied many of the European settlements across the region.

With the introduction of the Aboriginals Protection and Restriction of the Sale of Opium Act 1897, the Queensland State Government sought to standardize the administration of the Aboriginal population:

> attempting to apply in a new and systematic fashion the principle of management through rationing that had been unevenly and hesitantly practised for a generation. Aboriginal people not working for whites would be forced to congregate at rationing points that made up a state-wide "welfare" network.
>
> (Rowse 1987: 85)

Among the effects of the new legislation was the establishment of reserves, areas allotted for the housing of the Aboriginal population under the control of local Protectors. In Coen the senior local policeman held this post. The Protectors held extensive powers over their Aboriginal charges, including the arrangement of work contracts and marriages on their behalf and the control of their physical location and movements. Work contracts on stations were organized for several months at a time and, on their conclusion, older Coen people recall that they returned to the township for only a week or so before they were sent out on a new contract. Mixed-race children, and those not holding to work arrangements, committing crimes or behaving in an unsatisfactory manner could be removed to a reserve elsewhere on the Peninsula or further south. By the 1930s, the use of Aboriginal labour from the camps and the reserve became increasingly systematized, and the younger generations raised there were more fully incorporated into the new rural industries. In 1936, 95 per cent of station work was being carried out by indentured Aboriginal labour. The region's fringe camps were cleared and their populations moved to missions or into Coen, and the region was fundamentally transformed through Aboriginal encapsulation within the white colonial system. Even then, some fringe camps persisted; the last camp to be cleared, at Port Stewart on the east coast, was not removed until 1961.

Spatial organization now revolved around a series of population centres – the township of Coen, the cattle stations and their outstations – the mining camps having all but disappeared after World War II. Coen was the administrative and economic centre of the region and its peripheral Aboriginal reserve housed Aboriginal families and workers on their occasional return to the township. The cattle stations again housed Aboriginal women and children and in some cases older dependants, whilst men were away for several weeks at a time working cattle. To this end "outstation" camps were developed as satellites to the main station homestead, which provided secondary centres for cattle work. In some cases these became smaller homesteads, with their own ration-stores and female and infant populations. Station demography thus confined Aboriginal women and children, for the most part, to the main cattle camps, whilst men ranged the surrounding landscape of dams, yards, dinner camps and night paddocks, mustering cattle for branding or for droving to the south for sale at the base of the Peninsula (Smith 2000, 2002).

This pattern of life led to many men maintaining contact with a continuing Aboriginal landscape, "traditional" knowledge being passed on through travelling along mustering routes. "Holidays" in the bush during lay-off periods from cattle work additionally allowed families to range occasionally over the station area (Smith 2002; May 1994). Initially the coalescence of station populations from surrounding areas and the development of relationships between Aboriginal workers and white "bosses" allowed many people to remain on or near their own estates and the areas over which they and their forebears had previously ranged. However, over time an increasing amount of scattering was evident, with many people working for much or all of their careers on what had previously been "stranger country". Nonetheless, these men and women also developed an extensive knowledge of their new environment.

In later years, the station populations were increasingly narrowed down to working men and women and their children, with the population at the Coen Reserve growing commensurately. With the introduction of equal wages for Aboriginal workers and the payment of social security to Aboriginal people, Indigenous employment in the region's cattle industry collapsed in the late 1960s, and the region's Aboriginal population was concentrated on the reserve. This period is identified, by older Aboriginal people now living in the region, as the most disruptive one for Aboriginal life, and in particular for relations with and knowledge of "country".[6] The contemporary appearance of the region's outstation movement is said, by many of these people, to have come "thirty years too late". For these thirty years, the Coen Aboriginal population was, for the most part, confined to the township, dependent on welfare payments, with alcohol use prevalent and an increasingly disrupted social fabric emerging (Smith 2000).

But from early on in this period one or two of the region's more enterprising families – perhaps not coincidentally those most confident in their relationships and communication with the town's dominant white minority – gained access to vehicles and used them to re-visit country which held particular significance for them for short stays and visits. Sunlight

Bassani, the head of one of these families and a prominent man in the contemporary coastal Lamalama language-named group, recalls these visits, saying that:

> me and Florrie we went back in the 70s. And the kids. I used to hire car, he take me down and we spend time . . . that'll be around the 70s. I had my own car then in '75, some more (Lamalama people) come down and we all went down. Second one was Keith [Liddy, Florrie's brother] and his wife, we all went down there. Keith used to make a home there, he was working at Yarraden [Station]. When he had a month's holiday, he spent his month down Port Stewart instead of Coen.

> (Port Stewart Lamalama and Centre for
> Appropriate Technology 1997: 9–10)

Decentralization and self-determination

These ongoing visits to country and to places of particular importance were facilitated by access to vehicles, usually provided by sympathetic employers – typically station owners or the owners of local stores. They involved both "pull" and "push" factors; Sunlight and Florrie Bassani's family, for instance, having both worked (in Sunlight's case) and lived (in Florrie's case – her family had remained at their Port Stewart camp until it was removed in 1961) at Port Stewart, have maintained a deep connection of knowledge, feeling and ownership under Aboriginal law with the surrounding country. Holidays at Port Stewart also provided an escape from the increasingly stagnant and decaying social milieu of the town, where unemployment, alcohol and social disruption were taking their toll.

Available evidence suggests that the constraints on Aboriginal mobility this time were focused on availability of vehicles and other resources, and the lack of access, through work opportunities or otherwise, to country, most of which lay in surrounding pastoral leases. Certainly, with the formation of two Aboriginal corporations in the 1980s, the emergence of Aboriginal land-rights and a national push for Aboriginal "self-determination", these occasional visits formed the basis for the emergence of a broader movement of decentralizsation in the region.

The first such organization was the Malpa Kincha Corporation formed in 1984 with the assistance of the local Lutheran pastor, although Aboriginal people in Coen had been attempting to have a local Aboriginal representative organization established since at least 1978 (FA&IRA 1979: 187). The need for such a corporation appears to have been driven both by aspirations for local and regional political representation and a necessity to form an incorporated body to tap into government funding (Jolly 1997: 241–2). Malpa Kincha proved initially successful, representing all Coen groups, but inter-family tensions led to a number of families leaving the organization and forming a second body, the Moomba Aboriginal Corporation, in 1987. The division of the town's Aboriginal population between the two corporations, increasing tensions between family groups as land claims and outstation support became issues and a desire to establish a local CDEP (workfare) programme, which ATSIC (Aboriginal and Torres Strait Islander Commission), the Commonwealth Government's Aboriginal arm, refused to fund unless Coen was represented by a single body, all led to the planning of a new corporation. Initially this was to be named CYCAD (Cape York Central Aboriginal Development) Corporation, but it was derailed as local Aboriginal people valued their autonomous family group interests over the benefits that a single organization might accrue (see Jolly 1997: 246–56). The Coen Regional Aboriginal Corporation (CRAC) was finally incorporated in February 1993, with a board of directors which represented each of the Coen-based language-named groups, excepting one group which chose to be represented by Aurukun Community Incorporated (see Smith 2000), and began full operations in April of that year. A CDEP programme began operating two months later, with forty-seven participants, which grew quickly to around ninety participants over the following months (CRAC 1993: 3).

The resources available through the Corporation, alongside a series of land purchases and transfers and the pursuit of land claims under Queensland's Aboriginal Land Act 1991 and the Commonwealth's Native Title Act 1993, rapidly facilitated the development of "outstations", commonly understood to be "small decentralised communities of close kin established by the movement of Aboriginal people to land of social, cultural and economic significance to them".[7] The term itself comes from the pastoral industry, where it refers to the smaller satellite stations at a remove from a main station homestead settlement used in mustering vast pastoral properties.[8] For the Coen region I take an "outstation" (in the local Aboriginal context) to be a perennial or seasonally re-established camp or settlement, serviced and administrated by, but distinct from, a larger population centre with which it has strong social ties, including the occasional residence of its population. This population in turn consists of a smaller group of Aboriginal people than those commonly living in the settlement, the majority of whom hold and assert "traditional" and/or "historical" ties with the area in which the outstation is situated and close ties of kinship and socio-cultural identity with each other.

Across northern Australia, the "outstation" or "decentralization move-ment," in which smaller groups broke off temporarily, seasonally or semi-permanently to establish these smaller communities began in the early 1970s (Coombs 1974: 5–7; Brokensha and McGuigan 1977: 119–20; Gray 1977: 114–17; Coombs *et al.* 1982: 427–30), although – as with the Coen region – it is apparent that such "movements" had parallels in earlier "de facto" shifts in many cases (de Graaf 1977; Eriksen 1996). Its emergence was concurrent with the Federal Government-sponsored regaining of control of land and the availability of both government funds and cash in the form of welfare benefits to Aboriginal families and groups. A major breakthrough was the 1973 Commonwealth Labour Government's commitment to "support decentralization both in principle and in practice . . . [following which] the exodus began in earnest" (Coombs *et al.* 1982: 428). For Queensland, however, an assimilationist and interventionist State administration remained firmly opposed to land rights, self-determination and decentralization, with the result that, with rare exceptions, decentralization and the establishment of outstations was near impossible until the early 1990s (Eriksen 1996).

From the inception of this "movement", outstations have been a focus of considerable attention, including anthropological investigation, although most of the work with outstations as its direct focus has been policy or programme oriented.[9] Early non-policy responses to the "movement" include anthropological (e.g. Meehan and Jones 1980; Coombs *et al.* 1982; Borsboom 1986) and medical and psychiatric approaches (Morice 1976), as well as academic accounts by service providers and agency liaison officers (e.g. Brokensha and McGuigan 1977; de Graaf 1977; Gray 1977; Gerritsen 1982). Common emphases in these accounts are the benefits and improvements in wellbeing experienced by those making use of outstations and that the main force behind decentralization movements has been local Aboriginal aspirations. There also remains a common assumption that this "movement" is, to some extent, a homogenous one in terms of its motivations and bases. From a fieldwork-based perspective, this last assumption appears problematic and in need of informed revision. The different "traditional" cultures, forms of contemporary social change and bureaucracies across Australia necessarily generate very different foundations for and forms of social action, and it is apparent that "outstation movements" should prove no exception (see Davis and Arthur 1998: 1–2). Nonetheless, regional differences within what is commonly taken to be a single phenomenon remain grossly under-examined both in academic and policy environments.

With the formation of the Coen Regional Aboriginal Corporation and the start of the CDEP scheme, the resources and impetus for further outstation development were introduced to the Coen region. By 1994 the region had gained eight outstations with perennial, seasonal or aspirational populations (Cooke 1994a, b), with a further two appearing in 1995. During the period in which I first conducted fieldwork (1996–1997) all except one of these outstations were in regular use by smaller or larger groups. The Coen region outstations in use in 1995–1997 are shown in Table 12.1.

In most of these cases, a number of factors were associated with the development of the outstation and continuing outstation residency. Principal amongst these have been the role of focal men and women who act as "boss" for the outstation camp, and play a key role both among the extended family group associated with the camp, the wider regional

Table 12.1 Coen region outstations, 1995–1997

Outstation	Group	Associated area	Occupancy	Type of Tenure
Punthimu ('Station Creek')	Ayapathu	Directly south and east of Coen	Seasonal, tents	No formal tenure, on expired pastoral lease.
Moojeeba	Lamalama	Along coast and inland east and south-east of Coen	Perennial/ seasonal, sheds and tents	On land transferred under ALA[a].
Theethinji (two sites around 3 km apart at Port Stewart)			Seasonal, tents and sheds	
Merapah	'Merapah Mob' (Wik Iyeny)	'Tablelands' around 60 km west and south-west of Coen	Station housing	Aboriginal-run pastoral lease held in trust.
Langi	Mungkanhu	West and north-west of Coen	Tents	No formal tenure, in National Park
Wenlock	Northern Kaanju	Around 60 km north of Coen	Sheds	In DOGIT[b] (former mission) reserve
Mulundudji			Occasional camping	On DPI[c] managed state land
Glen Garland	Olkola	Around 80 km south of Coen	Station housing	Aboriginal-run pastoral lease held in trust.
Birthday Mountain	Southern Kaanju	Directly north and northeast of Coen	Tents and sheds	On land claimed under ALA.
Stoney Creek			Tents and sheds	In DOGIT (former mission) reserve

Source: Modified after Smith 2000: 66.

Notes:
[a] Aboriginal Land Act
[b] Deed of Grant in Trust
[c] Department of Primary Industries.

population and in relationships with regional and governmental organizations in gaining financial and other resources and support for the ongoing
existence of the outstation. The same men and women play key roles in
land claims, which, although not vital to the establishment and use of all
the region's outstations, have regenerated a situation in which the
Aboriginal families of the region expect to regain control of their "traditional land" and consider themselves as having the right to visit and use
their country despite its current tenure situation. Another key factor in the
role of these focal men and women is their control of vehicles – at the time
of my initial fieldwork mostly gained through government funding, but
increasingly privately owned – through which they facilitated their own
visits to the outstation and town and other settlements in the region, and
the movements of their kin (Smith 2000).

Contemporary mobility patterns

In the contemporary period, the access to and use of vehicles, the establishment of outstation camps, the dispersal of kin across the wider Peninsula
region and the development of ties to other "communities" which are
commonly visited during important events such as funerals or sporting
carnivals (see Young and Doohan 1989) are all associated with a renaissance
in regional population mobility. The greatest single form of mobility is
movement between outstation camps and the township of Coen. The levels
of mobility are illustrated in Tables 12.2 and 12.3, which show the mean
weekly mobility of the wider "tribal" groups associated with four of the
region's outstations in the dry and wet seasons of 1996 and 1997.

It is clear from the tables that the wider Coen population – most if not
all of whom are identified more or less with a particular language-named
outstation ("tribal") group – move frequently between the outstation with
which they are identified, and often to other outstations, Aboriginal
communities and the regional centre of Cairns, some 700 km to the south.
Moreover, there is a clear seasonal shift in mobility patterns, with outstation
mobility increasing in frequency in the dry season, and a more sedentary
Coen-based lifestyle becoming evident in the wet season when some
outstation camps are occasionally inaccessible by road because of inundated
river crossings or boggy plains. Whatever the direction of relocation, the
majority of population movements occur within the Coen region throughout the year.

The reasons for contemporary mobility patterns are complex and
involve a series of push and pull factors, these factors demonstrating the
continuity and simultaneous transformation of many factors apparent in
pre-colonial mobility practice interwoven with a series of new forms and
motivations. Thus, movement to outstations involves the desire to spend
time on one's own country or that of affines in the company of a small
group of close relations, to access bush-foods, materials and other resources

Table 12.2 Dry season mobility patterns (percentage), 1996–1997

	Moojeeba	Theethinji	Wenlock River	Langi
At the outstation all week	11	17	5	8
At outstation some of week, moved (between Coen and outstation) once	26	18	17	18
At outstation some of week, moved (between Coen and outstation) twice or more	23	12	27	21
In Coen all week	13	30	17	22
Moved in Coen region, but not to or from own group's outstation	18	18	25	22
Outside of Coen region all week	9	5	9	9
Total	100	100	100	100

Source: Modified after Smith 2000.

Table 12.3 Wet season mobility patterns (percentage), 1996–1997

	Moojeeba	Theethinji	Wenlock River	Langi
At the outstation all week	16	21	0	0
At outstation some of week, moved (between Coen and outstation) once	22	13	3	0
At outstation some of week, moved (between Coen and outstation) twice or more	14	3	8	0
In Coen all week	34	51	46	43
Moved in Coen region, but ot to or from group's outstation	10	12	35	40
Outside of Coen region all week	4	0	8	17
Total	100	100	100	100

Source: Modified after Smith 2000.

and a desire to escape the noise and tensions of life in the township. For some – focal men and women and members of their families acting as drivers of outstation vehicles – it is also tied to obligations to kin, and to move other people and resources, including shopping from the township to those already at the outstation.

Similarly, movement into Coen is linked to boredom or restlessness with the relative quiet of bush life and the desire or need to go into town to shop, gamble and drink, to take bush resources – in particular, bullock meat, turtles and fish – to distribute to kin there, attend meetings, health clinics or school and to transport others into town for these same reasons. The weekly and fortnightly round of CDEP pay, pension and welfare cheques and doctor's visits to the township's clinic has a strongly structuring effect on people's mobility, these events seeing a strong tendency to shift to Coen, the regional administrative centre, to collect cheques, spend money and meet appointments.

The motivations for movements differ predictably with age, gender, social status and personality. For some younger people the desire to move is based firmly in what N. Peterson (personal communication) has called "existential mobility", particularly among those younger men whom von Sturmer refers to as "floaters" (von Sturmer 1978; see also Martin and Taylor 1996) – socially peripheral individuals who move repeatedly both within and between the places in which they are usually short-term residents. The movements of such individuals are perhaps best seen as part of a spectrum of younger people's movements. In most cases at a remove from substantial responsibilities, and with ties to two or more of the region's outstations (residence at the outstation of one's mother's tribal group is as common as at one's father's mob's outstation), many young people take the opportunity to move between a set of places. Such patterns of mobility open up a wider selection of people, foodstuffs and activities than would otherwise be the case, lessening what many younger Aboriginal people see as the boredom of Coen life. Whilst "floaters" are typically more peripheral, exploiting relatively weak kin ties for shelter and resources, younger people in general are able to exploit the opportunities of nebulous social inclusion by "floating" from place to place, partly in an attempt to redress the boredom associated with particularly pronounced exclusion from economic participation and meaningful social roles.

A propensity and desire for movement is apparent across the wider population, although there is a need to further differentiate the movements of focal men and women, whose mobility is frequent but tied to more specific personal and group-linked purposes. It is not uncommon for a car or truck-load of people from another settlement to pull up at a relative's house in town, or at an outstation camp and call out to kin trying to persuade them to jump on, these new passengers often leaving without even a bed-roll or a change of clothes, departing to another location and returning weeks or months later. In the meantime frequent phone calls will be made to family, the post office and administrative staff to arrange for money and bank books to be sent to them, or for a flight to Coen (or to another community when visitors found themselves stuck in Coen) to be organized and paid for. Such mobility, of course, is dependent on a network of kin ties across a number of settlements and camps in the region. People are far less likely to jump onto a truck heading somewhere where they have few close ties and where, as a result, they will not be able to be looked after by kin, affines or friends. Similarly, Coen people visiting the city of Cairns will choose to stay with white friends, typically those employed in regional service provision agencies, rather than stay in motels where the choice is available.[10]

Other classical patterns and motivations for population mobility demonstrate clear parallels with the situation at the threshold of colonization. Mobility to avoid conflict is common, with inter-personal tensions, fights and other forms of violence commonly provoking a shift from Coen to outstation camps or to other communities in the wider region. Likewise, although some of the structuring factors are different (restricted road access as opposed to changes in resource availability for example), the seasonal emphasis on smaller bush camps or outstations in the dry season and larger supercamps (McKnight 1986) in the wet season continue a "classical" pattern of population mobility, and demonstrate the same late-wet-season desire to escape the larger population conglomerations as soon as possible after several months of psychological strain, inter-personal tensions and fights.

Mobility, policy and the state

A general notion surrounding outstation or homeland movements has been that they are "return to country movements", and that they mark an attempt to move away from the population centres established during the colonial period and move back to traditional country. Thus, in an early review of the phenomenon of decentralization, Coombs *et al.* (1982: 428) write of the "exodus" from regional centres which began following the Commonwealth Government's announcement of support for decentralization. Recent accounts of decentralization from both academia and the policy arena similarly tend to stress decentralization as a "return to country", a description that, although partially accurate in that such movements re-invigorate Aboriginal people's involvement with particular places and areas with which they have a "traditional" connection, mask the ongoing ties with regional centres that remain apparent in the Coen region as elsewhere (Smith 2000; Peterson 2000).[11]

This emphasis leads not only to dangers of apprehension of the nature of decentralization and outstation movements, but to policy decisions which can undermine Aboriginal aspirations or set people up for "failure" in terms of perceived development outcomes. The Aboriginal and Torres Strait Islander Commission's[12] 1996 report on *Community Infrastructure of Homelands* (ATSIC 1996), for example, recommends funding outstations relative to the periodicity of their use, potentially undermining the use and development of outstations for which seasonal or occasional use is normative and from which benefits in areas such as Indigenous health and transmission of cultural knowledge may accrue. Similarly, a development report focused on an outstation in the Coen region emphasized that the members of the associated language-named group, who lived at the outstation, in Coen and a number of other settlements across the Peninsula would increasingly return to live at their "homeland" in response to funding of infrastructure there. Although there is no doubt that, at least in the short term, and perhaps over a longer period, such development has heightened the levels of visits to the outstation, most of the locally based population have continued to maintain a bi-locational residence and move between the outstation and Coen, and little or no difference has been seen in the visits of those living elsewhere, most of whom have strong ties of residence and sociality to the communities in which they now reside. In this case, an emphasis on "exodus" to the outstation and the surrounding homeland appears to have produced a flow of funding and resources on the basis of a demographic transformation that has not occurred, endangering further support for this outstation and others across the region.[13]

Likewise, recommendations by a regional research and community development project that funding agencies "should not fund 4wd [four-wheel drive] vehicles for individual groups to meet transport needs between outstations and service centres on outstations" (Cooke 1994a: 43) appear to have been taken up at both state and national levels, with the subsequent restriction of availability of funding for vehicles severely impinging on the ability of several groups in the Coen region to move between their outstation and Coen, vehicle access and use being a key aspect of population mobility in the Coen region. Not only were these groups' aspirations for outstation use and development hampered by the lack of access to vehicles, but the resulting frustrations had clear impacts on group members including a resumption of binge drinking among those whose outstation involvement had led to a marked decrease in substance abuse.[14]

Despite an insistence – both in the literature and among many Aboriginal people themselves – that decentralization and regenerated population mobility allow Aboriginal people to remove themselves from both enforced habitation alongside other Aboriginal groups and the assimilatory and acculturating effects of state domination in the township of Coen, the underpinnings of the outstation movement are in fact deeply inter-ethnic, both in the sense of a regional Aboriginal "community" and through the interpenetration of the "Aboriginal" and "whitefella" domains across the region (see Smith 2000; Dagmar 1990). The policy environment in which "self-determination" and "land rights" have meant access to land, vehicles and other resources has been the foundation of the outstation movement, and this situation has arisen from the particularly favourable grounds for Aboriginal aspirations available in liberal democratic nation-states (Peterson 1998, 1999). Similarly, in the Coen region, it has been the conjoint action of a number of different extended family or language-named ("tribal") groups – themselves recognizable as "ethnically distinct"[15]

– which has allowed the development of the self-determined development, focused on outstations, of each of these groups, in particular through their co-operation through the Coen Regional Aboriginal Corporation. In both cases the outstation movement underscores a contemporary version of the articulation of "autonomy and relatedness" which has long been a principal core to Aboriginal life (Myers 1986; Martin 1993), but which now encompasses both a wider regional Aboriginal polity (see Peterson 2000) and relationships with local whites, regional organizations and state agencies.

In common with other decentralization, outstation or homelands movements, the main force behind decentralization in the Coen region has been local Aboriginal aspirations. Elsewhere, drawing on a term used by Berndt (1962) and by Wilson (1979), albeit in different contexts, I have described decentralization in the Coen region as an "adjustment movement" in which valued practices are able to continue, new practices and resources are accommodated, underlying structures are maintained in cultural production and the status of key Aboriginal men and women is maintained. Here, unlike the Arnhem Land situation to which Ronald Berndt originally applied the term, the adjustment movement is not only a shift in social life, but one manifested through population mobility. Outstations and those who use them are both, in their own ways, "between places" – that is, people continue to move between outstations and Coen, and the outstations themselves are an interim measure, initiated by local Aboriginal people in their attempts to adjust to changing post-colonial circumstances (Smith 2000). Rather than an "exodus", decentralization adjusts what Kolig (1978) has called Aboriginal life-spaces so that they are able to encompass a range of places, populations and resources. The importance of contemporary population mobility is its role in mediating these dimensions of persistence and change (Merlan 1998; see also Chapter 2) within Aboriginal lives.

Simultaneously, the forms of support and involvement of various aspects of the state and of local and regional Aboriginal organizations are also aimed – intentionally or otherwise – at making adjustments to the contemporary situation, society and economy in which Aboriginal people find themselves. The outstation movement is a process (Young and Doohan 1989) in which different agencies and aspirations articulate, and one in which Aboriginal people, through mobility, have asserted their own values, intentions and aspirations. As Taylor and Bell note in Chapter 2, unlike other situations where community development leads inexorably towards socio-cultural convergence and urbanization, the particular features of local Aboriginal life, in combination with encapsulation by a liberal democratic polity and welfare state, have allowed a form of articulation in which forms of difference based in social and spatial separation have continued, albeit in forms that are continually changing through their local and wider articulations with other agencies. The importance of mobility in managing the social, political and economic lives of the Aboriginal people of central

Cape York Peninsula has remained, despite the broadened and radically different milieu in which today's Aboriginal population find themselves.

The outstation movement is not a "return to the past" or to "tradition", but the ongoing manifestation of local values and forms in articulation with a state and "mainstream" socio-cultural milieu which is at times compatible with and at other times in opposition to local Aboriginal life-worlds. Through decentralization Aboriginal people – and particularly a set of motivated and visionary men and women embedded in what remains seriously damaged social fabric – have sought an alternative to assimilation and alienation through re-invigorating their involvement with places of social and economic importance. They have thus maintained their ties to the township, regional centres and "traditional country", and mediated their involvement with these places and the agencies which are manifested in them through the ongoing practices of population mobility.

Notes

1 The research on which this chapter is based took place mostly in and around Coen, Far North Queensland in 1996 to 2002, and was supported by a Study Abroad Studentship from the Leverhulme Trust, a Research Grant from the Australian Institute of Aboriginal and Torres Strait Islander Studies, the Emslie Horniman Fund of the Royal Anthropological Institute, the University of London Research Fund and the Australian Research Council. I am indebted to the Aboriginal people of the Coen region in writing this paper, and in particular to Phillip Port, Robert Nelson, Vera Claudie, David Claudie, Margaret Sellars, Sunlight and Florrie Bassani, Victor Lawrence and Ann Creek, for their support for my work and their patience with my more than occasional lack of understanding. This research has benefited from numerous discussions with colleagues, notably John Taylor, Nic Peterson, Bruce Rigsby, David Martin, Mark Moran, David McKnight, Ray Wood, Lachlan Walker and Su Groome. I am also grateful for the insights and assistance offered by Caroline MacDonald, Chris Bradley and Annalies Voorthius at the Coen Regional Aboriginal Corporation during my research.

2 For example, the evidence for the rainforest region around Cairns *c*. 1900 points strongly to relatively permanent villages of lawyer cane or grass huts with numerous seasonal camps up and down the river systems (R. Wood, personal communication). Dixon (1972) notes that the Dyirbal-speakers of the Cairns–Tully rainforest region "moved camp quite frequently in the dry season, building low sleeping huts as they were needed" but that "[d]uring the heavy rainfall of the wet season they would return to larger, more permanent huts (that might last for several years)" (Dixon 1972: 28; see also Peterson and Long 1986: 44). Interestingly, there are economic similarities (fish-traps, eel-cooking structures) between this region and Gunditjmara country around Lake Condah in western Victoria, where "cooking trees" for smoking eels, fish-traps and stone huts also point to more settled communities – estimated at up to 600 people by Heather Builth, an archaeologist working in the area – than generally supposed (*Weekend Australian* 2002). Donald Thomson notes of Cape York Peninsula (apparently referring both to Wik and northern Princess Charlotte Bay peoples) that

> people may spend several months of the year as nomadic hunters . . . and exploiting the resources of vegetable foods . . . A few months later the same

people may be found established on the sea coast in camps that have all the appearance of permanence or at least of semi-permanence, having apparently abandoned their nomadic habits. They will remain in these camps for months on end . . . viewing these independently at different periods of the year, and seeing the people engaged in occupations so diverse, an onlooker might be pardoned for concluding that they were different peoples.

(Thomson 1939: 209; see also Peterson and Long 1986: 43–4)

3 I use this term following the anthropological convention of describing the economic organisation of some Indigenous peoples in Papua New Guinea, South East Asia, etc. as that of "shifting cultivators". Elsewhere (e.g. Smith 2000) I have employed the term "semi-nomadic".

4 These were not local groups, instead forming the population of the "bands" ranging across their own, or others' clan estates. Peterson and Long (1986) have articulated a complex mechanism of relatedness between these land-owning and land-using groups. See also Sutton (1999, 2001) and Thomson (1939: 211, fn.1).

5 In fact, the majority of the region's Aboriginal outstations are located at or near the sites of former cattle or mining camps (see Smith 2000, 2002).

6 "Country" is land as property within an Aboriginal system of land tenure but, rather than being a passive object, country is also seen as possessing agency (see also Arthur 1996: 120–1).

7 Commonwealth of Australia (1987: 7) cited in Davis and Arthur (1998: 2).

8 Other terms, notably "homelands" or "homeland centres", are also used to designate these decentralized Aboriginal communities. Many (e.g. Altman and Taylor 1989) use the terms "outstation" and "homeland (centre)" interchangeably. "Outstation" is both the common term used by Indigenous people and organizations in the Coen region and indicative of the relationship between contemporary decentralization and the region's cattle industry. Davis and Arthur (1998: 2–4) discuss the use of the term "homelands" in some detail.

9 As Davis and Arthur (1998: 1) note, this may be particularly true of more recent years, where the topic has been "talked out" in the public domain but has continued to develop in communities themselves, particularly with the establishment of ATSIC.

10 A second group of Aboriginal people from the region live for shorter or longer terms as "park drinkers", sleeping rough in a number of locations in and around Cairns.

11 In part, such notions of "exodus" reflect widespread fantasies among sympathetic non-Indigenous people of a "return to country" that is a return to tradition and a more "authentic" Aboriginal existence (see Smith 2000).

12 The Aboriginal and Torres Strait Islander Commission (ATSIC) is a Commonwealth statutory authority, elected by Indigenous people as Australia's peak representative body for Aboriginal and Torres Strait Islander interests.

13 It may be that the definition of the consultancy steered the consultants involved in the planning process to predict such a transformation, although staff at the Coen Regional Aboriginal Corporation and members of the outstation group involved with the project both stressed that such an "exodus" was unlikely to occur. At public meetings with the consultants and representatives of other agencies, the focal man of the outstation group involved in the project found himself in the impossible situation of not wishing to contradict the bases on which his group might receive much-needed resources, whilst also not wishing to agree to impossible predictions of population increases at the outstation over the following ten years. He deferred in the face of these assertions, stating that this was a "hard question".

258   *Benjamin Richard Smith*

14 See Brady (1991), Martin (1998) and Pearson (2000) on substance abuse in Aboriginal communities in Australia.
15 Here I follow Barth's proposed definition of an ethnic group as (i) largely biologically perpetuating, (ii) sharing fundamental cultural values, realized in overt unity in cultural forms, (iii) making up a field of communication and interaction and (iv) having a membership which identifies itself, and is identified by others, as constituting a category distinguishable from others of the same order (Barth 1969; see also Chase (1980) on "ethnicity" and Aboriginal people at Lockhart River; Smith 2000: 416–17).

References

Altman, J. C. and Taylor, L. (1989) *The Economic Viability of Aboriginal Outstations and Homelands*, Canberra: Australian Government Publishing Service.
Arthur, J.M. (1996) *Aboriginal English: A Cultural Study*, Melbourne: Oxford University Press.
ATSIC (1996) *Community Infrastructure of Homelands: Towards a National Framework*, ATSIC Discussion Paper, Canberra: ATSIC.
Barth, F. (1969) 'Introduction,' in F. Barth (ed.) *Ethnic Groups and Boundaries: The Social Organization of Culture Differences*, London: George Allen & Unwin.
Berndt, R.M. (1962) *An adjustment movement in Arnhem Land, Northern Territory of Australia*, Paris: Mouton.
Borsboom, A.P. (1986) 'The cultural dimension of change: an Australian example,' *Anthropos*, 81: 605–15.
Brady, M. (1991) 'Drug and alcohol use among Aboriginal people,' in J. Reid and P. Trompf (eds) *The Health of Aboriginal Australia*, Sydney: Harcourt Brace Jovanovich.
Brokensha, P. and McGuigan, C. (1977) 'Listen to the Dreaming: the Aboriginal homelands movement,' *Australian Natural History*, 19(4): 118–23.
Chase, A. K. (1980) 'Which way now? Tradition, continuity and change in a North Queensland Aboriginal community', unpublished Ph.D. Thesis, Department of Anthropology and Sociology, University of Queensland.
Chase, A. K. (1984) 'Belonging to country: territory, identity and environment in Cape York Peninsula, Northern Australia,' in L.R. Hiatt (ed.) *Aboriginal Landowners: Contemporary issues in the Determination of Traditional Aboriginal Land Ownership*, Oceania Monograph 27, Sydney: University of Sydney.
Chase, A.K. and Sutton, P. (1998/1981) 'Australian Aborigines in a rich environment,' in W.H. Edwards (ed.) *Traditional Aboriginal Society* (2nd edition), Melbourne: Macmillan.
Commonwealth of Australia (C.A. Blanchard, Chair) (1987) *Return to Country: The Aboriginal Homelands Movement in Australia*, Report of the House of Representatives Standing Committee on Aboriginal Affairs, Canberra: Australian Government Publishing Service.
Cooke, P. (1994a) *Planning a Future at Cape York Peninsula Outstations*, Cairns: Cape York Land Council.
Cooke, P. (1994b) *Cape York Peninsula Outstation Strategy: Summary, Recommendations and Proposed Action*, Cairns: Cape York Land Council.
Coombs, H. C. (1974) 'Decentralisation trends among Aboriginal communities,' *Western Australia Aboriginal Affairs Newsletter*, 1(8): 4–25.

Coombs, H.C., Dexter, B.G. and Hiatt, L.R. (1982) 'The outstation movement in Aboriginal Australia,' in E. Leacock and R. Lea (eds) *Politics and History in Band Societies*, Cambridge: Cambridge University Press.

CRAC (Coen Regional Aboriginal Corporation) (1993) *Coen Regional Aboriginal Corporation Community Development Plan*, Coen: Coen Regional Aboriginal Corporation.

Dagmar, H. (1990) 'Development and politics in an interethnic field: Aboriginal interest associations,' in R. Tonkinson and M. Howard (eds) *Going it Alone? Prospects for Aboriginal Autonomy*, Canberra: Aboriginal Studies Press.

Davis, R. and Arthur, W. S. (1998) *Homelands and Resource Agencies Since the Blanchard Report: A Review of the Literature and an Annotated Bibliography*, CAEPR Discussion Paper 165, Canberra: Centre for Aboriginal Economic Policy Research.

de Graaf, M. (1977) 'Some aspects of decentralisation in Aboriginal communities in Central and Western Australia,' *Readings in Bilingual Education*, 84.

Dixon, R.M.W. (1972) *The Dyirbal language of North Queensland*, Cambridge Studies in Linguistics 9, Cambridge: Cambridge University Press.

Eriksen, J.P. (1996) 'Aboriginal outstation development in Cape York Peninsula: moving back to traditional country', unpublished Honours Thesis, Division of Environmental Sciences, Griffith University, Brisbane: Griffith University.

FA&IRA (Foundation for Aboriginal and Islander Research Action) (1979) *Beyond the Act*, Brisbane: Foundation for Aboriginal and Islander Research Action.

Gerritsen, R. (1982) 'Outstations: differing interpretations and policy implications,' in P. Loveday (ed.) *Service Delivery to Outstations*, Darwin: North Australia Research Unit, The Australian National University.

Gray, W. J. (1977) 'Decentralisation trends in Arnhem Land,' in R.M. Berndt (ed.) *Aborigines and Change: Australia in the '70s*, Canberra: Australian Institute of Aboriginal Studies.

Jolly, L. (1997) 'Hearth and country: the bases of women's power in an Aboriginal community on Cape York Peninsula', unpublished Ph.D. Thesis, Department of Anthropology and Sociology, University of Queensland.

Keen, I. (2001) 'Variation in Indigenous economy and society at the threshold of colonisation', unpublished paper presented at *The Power of Knowledge, the Resonance of Tradition – Indigenous Studies: Conference 2001*, Manning Clark Centre, The Australian National University, Canberra, 18 to 20 September 2001.

Kolig, E. (1978) 'Dialectics of Aboriginal life-space,' in M.C. Howard (ed.) *'Whitefella Business': Aborigines in Australian politics*, Philadelphia: Institute for the Study of Human Issues.

Loos, N. (1982) *Invasion and Resistance: Aboriginal–European Relations on the North Queensland Frontier 1861–1897*, Canberra: Australian National University Press.

McKnight, D. (1986) 'Fighting in an Australian Aboriginal supercamp,' in D. Riches (ed.) *The Anthropology of Violence*, Oxford: Basil Blackwell.

Martin, D.F. (1993) 'Autonomy and relatedness: an ethnography of the Wik people of Western Cape York Peninsula', unpublished Ph.D. Thesis, Canberra, Australian National University.

Martin, D.F. (1998) 'The supply of alcohol in remote Aboriginal communities: Potential policy directions from Cape York', *CAEPR Discussion Paper 162*, Canberra: Centre for Aboriginal Economic Policy Research.

Martin, D. and Taylor, J. (1996) 'Ethnographic perspectives on the enumeration of Aboriginal people in remote Australia', *Journal of the Australian Population Association*, 13(1): 17–33.

May, D. (1994) *Aboriginal Labour and the Cattle Industry: Queensland from White Settlement to the Present*, Cambridge: Cambridge University Press.

Meehan, B. and Jones, R. (1980) 'The outstation movement and hints of a white backlash,' in R. Jones (ed.) *Northern Australia: Options and Implications*, Canberra: Research School of Pacific and Asian Studies, The Australian National University.

Merlan, F. (1998) *Caging the Rainbow: Places, Politics, and Aborigines in a North Australian Town*, Honolulu: University of Hawai'i Press.

Morice, R.D. (1976) 'Woman dancing dreaming: psychosocial benefits of the Aboriginal outstation movement,' *Medical Journal of Australia*, 1976(2): 939–42.

Myers, F. (1986) *Pintupi Country, Pintupi Self: Sentiment, Place, and Politics among Western Desert Aborigines*, Oxford: University of California Press.

Pearson, N. (2000) *Our Right to Take Responsibility*, Cairns: Noel Pearson and Associates.

Peterson, N. (1998) 'Welfare colonialism and citizenship: politics, economics and agency,' in N. Peterson and W. Sanders (eds) *Citizenship and Indigenous Australians: Changing Conceptions and Possibilities*, Cambridge: Cambridge University Press.

Peterson, N. (1999) 'Hunter-gatherers in First World nation states: bringing anthropology home,' *Bulletin of National Museum of Ethnology*, 23(4): 847–61.

Peterson, N. (2000) 'An expanding Aboriginal domain: mobility and the initiation journey,' *Oceania*, 70(3): 205–18.

Peterson, N. and J. Long (1986) *Australian Territorial Organization: A Band Perspective*, Oceania Monograph 30, Sydney: University of Sydney.

Port Stewart Lamalama and Centre for Appropriate Technology (1997) *Moojeeba-Theethinji: Planning for a Healthy Growing Community*, Cairns: Centre for Appropriate Technology.

Rigsby, B. (1998) 'A survey of property theory and tenure types,' in N. Peterson and B. Rigsby (eds) *Customary Marine Tenure in Australia*, Oceania Monograph 48, Sydney: University of Sydney.

Rigsby, B. (1999) 'Genealogies, kinship and local group organisation: Old Yintjingga (Port Stewart) in the late 1920s,' in J.D. Finlayson, B. Rigsby and H.J. Bek (eds) *Connections in Native Title: Genealogies, Kinship and Groups*, Canberra: Centre for Aboriginal Economic Policy Research, The Australian National University.

Rowse, T. (1987) '"Were you ever savages?' Aboriginal insiders and pastoralists' patronage,' *Oceania*, 58(1): 81–99.

Smith, B. R. (2000) 'Between places: Aboriginal decentralisation, mobility and territoriality in the region of Coen, Cape York Peninsula (Australia)', unpublished Ph.D. Thesis, Department of Anthropology, London School of Economics and Political Science.

Smith, B. R. (2002) 'Pastoralism, land and Aboriginal existence in Central Cape York Peninsula,' *Anthropology in Action (UK)*, 9(1): 21–30.

Sutton, P. (1999) 'The system as it was straining to become? Fluidity, stability and Aboriginal country groups,' in J.D. Finlayson, B. Rigsby and H.J. Bek (eds) *Connections in Native Title: Genealogies, Kinship and Groups*, Canberra: Centre for Aboriginal Economic Policy Research, The Australian National University.

Sutton, P. (2001) *Aboriginal Country Groups and the Community of Native Title Holders*, National Native Title Tribunal Occasional Paper 1/2001, Perth: National Native Title Tribunal. Available online: http://www.nntt.gov.au (September 2001)

Thomson, D.F. (1934) 'The dugong hunters of Cape York Peninsula,' *Journal of the Royal Anthropological Institute* (OS), 64: 237–62.

Thomson, D.F. (1939) 'The seasonal factor in human culture: illustrated from the life of a contemporary nomadic group,' *Proceedings of the Prehistoric Society*, 5: 209–21.

von Sturmer, J. (1978) 'The Wik region: economy, territoriality and totemism in Western Cape York Peninsula, North Queensland', unpublished Ph.D. Thesis, Department of Anthropology and Sociology, University of Queensland.

Weekend Australian (2002) 'Rethinking the noble savage,' *Weekend Australian (Weekend Inquirer Section)*, 3–4 August, 2002.

Wilson, J. (1979) 'The pilbarra Aboriginal social movement: an outline of its significance,' in R.M. Berndt and C.H. Berndt (eds) *Aborigines of the West: Their Past and Present*, Nedlands: University of Western Australia Press.

Young, E. and Doohan, K. (1989) *Mobility for Survival: A Process Analysis of Aboriginal Population Movement in Central Australia*, Darwin: North Australia Research Unit, The Australian National University.

13 Conclusion

Emerging research themes

Martin Bell and John Taylor

This consolidation of groundbreaking research on Indigenous peoples' mobility in Australasia and North America provides a framework for eliciting emergent research themes. To this end, we are guided by Skeldon's (1997: 17–40) global exploration of the links between migration and development in which he identifies several theoretical and conceptual approaches to analysis. These he labels as economic approaches, studies of diaspora, transition theory, and postmodernist views. In this schema he identifies a tendency to shift over time from simple, if rigorous, economic models, towards more complex, but more subjective approaches. Also noted is a clear demarcation in approach between those studies focused on internal and those focused on international population movements.

While recent treatises on mobility research in developed countries have called for greater emphasis to be placed on biographical techniques (Halfacree and Boyle 1993), such approaches have a long tradition in studies of mobility among Indigenous peoples, as indeed they do in the developing world (Skeldon 1995). In the New World context, attention to local contingency reflects the longstanding dominance of anthropological research in the study of Indigenous populations, initially as an integral part of the colonial project. Accordingly, a wealth of ethnographic, community-based studies provides much of the cultural and experiential evidence of group and individual mobility.

Among the insights underscored by these ethnographic analyses is a clear reminder that Indigenous populations have homelands that are encapsulated by settler societies, and that the structural situations facing Indigenous peoples in the twenty-first century are very diverse. While many remain within their traditional lands, the story of much migration elicited through ethnographic texts is one of dislocation, of populations uprooted and compelled, either by force of might or circumstance, to take up residence in "exile". To this end, much of the population redistribution and subsequent related mobility that is described in this volume fits well within the concept of diaspora. This term, which evokes the dispersal of population groups, refers in particular to non-voluntary flight from homelands. Using Skeldon's (1997: 28–9) summary:

Diaspora draws attention to the migrant as victim, not only in the homeland but also at the destination, where minority communities are established. In destination areas, isolation, separate identity and discrimination are more recurrent themes than integration or assimilation . . . diaspora emphasises roots, exile and home. Irrespective of whether the migrants are victims or pioneers, central to the diaspora approach is the focus on networks: within the communities of destination, within those organizations that maintain group identity, and between origin and destination areas. Migration rather than a move for settlement is, from the point of view of diasporas, a transnational system of circulation.

Such themes are invoked in all the chapters in this volume. To the extent that aspects of Indigenous mobility reflect the characteristics of diaspora, then one important departure from the conventional conceptualization that requires theoretical development is that such movement occurs *within*, rather than *between*, modern nation states. This is internal migration, not international migration (except in the case of overseas Maori). Political scientists are well aware of the analytical and policy implications of colonial and post-colonial relations – of North American recognition (via treaties) of Indian Nations and First Nations, of New Zealand's Treaty of Waitangi and the reconstitution of Iwi, of Australia's Mabo decision and subsequent recognition of native title (Ivison *et al.* 2000; Nietschmann, 1994). Accordingly, they have grappled persistently with notions of "internal colonialism" in settler states.

By contrast, the demographic implications of internal colonialism and of diversity of circumstance have barely been posed, yet there is sufficient in the present volume to signal the relevance of such a conceptual framework for the analysis of mobility and migration. The connections are readily apparent: in the continuity of rural settlement and increased population dispersion on Indigenous lands, in customary forms of circulation, in return migration to Iwi, in population churning between reservations and cities, in historic data on forced exile, and in the policy implications of movement across jurisdictional boundaries and the related consequences of mobility for sovereignty or assimilation. In Australia, "diaspora" is already part of the lexicon of Indigenous mobility in the form of Torres Strait Islanders dispersed across the Australian mainland (Rowse 2002: 193).

Not surprisingly, then, the contributions to this volume stress the primacy of a political economy approach to mobility analysis. They expose the weakness of economic models that portray migrants as income maximizers in undifferentiated labour markets. This is not to deny that job search and depressed economic conditions in rural communities have been significant stimulants to post-war urbanization, but even here emphasis is placed more on the strength of government agency in directing or enabling migration flows, rather than on individuals as free agents. Alternative economic approaches that relate mobility to risk minimization within highly segmented labour markets present a more realistic framework for Indigenous populations, as they highlight the distinctiveness of Indigenous economic participation, mostly in secondary labour markets, and give prominence to the role of social networks in facilitating movement. Given the relatively low socioeconomic status of Indigenous populations in all four countries, however, the relationship between marginalization and mobility is a key question and one that requires much closer attention. Does Indigenous mobility reflect socioeconomic status, or does socioeconomic status reflect mobility?

At the same time, it is ironic that the remote location of many Indigenous groups on Indigenous-owned lands has resulted in growing and direct interaction with global economic forces, most strikingly in the form of mining companies and their related activities, and especially in Australia, Canada and the United States. In some instances, this has stimulated labour migration to large scale projects, but negotiated royalty and rental proceeds have also assisted in supporting dispersed settlement and more traditional pursuits.

The pivotal role played by large scale, structural forces – political, social, economic, technological – in shaping patterns of Indigenous mobility and settlement inevitably invites theorizing on a grand scale. Despite its manifest limitations, transition theory provides a powerful framework against which to examine large scale continuities and shifts in mobility across the broad sweep of history. There is a seductive elegance in the application of Zelinsky's (1971) hypothesis to mobility among Indigenous peoples, because many of the stimuli to their movement can be traced unambiguously to the impacts of settlement by the very non-indigenes whose migrations Zelinsky's original model sought to explain. Indeed, there appear to be fundamental connections between the shifting geography of Indigenous settlement and the processes of international migration and frontier expansion which form key elements of Zelinsky's (European) transition. Moreover, the search for an Indigenous variant is entirely consistent with Zelinsky's (1993: 218) own suggestion of a sizeable family of transitions and his call for blueprints of their sequential phases. Numerous parallels can be found between historical phases of the Indigenous settlement experience across the four New World countries examined here, as the comparisons in Part 1 of this volume clearly show. With the possible exception of New Zealand Maori, equally apparent is that the trajectory of Indigenous mobility is qualitatively different from the variants previously described for either the developed or developing worlds (Zelinsky 1971; Skeldon 1990), not only in spatial outcomes but also in the dynamics of population movement.

While transition theory concisely captures key shifts and phases in the evolution of settlement dynamics, mobility outcomes cannot be explained solely as the product of confrontation between large scale structural forces. As the papers in Part 3 reveal, they also reflect the subtle interplay with culture and tradition that mould spatial behaviour to diverse circumstances and geographical settings. As elsewhere, it is this tension between the global and the local, between the individual and the group, between culture and modernity, and ultimately between space and time, that holds the key to understanding of mobility among Indigenous peoples.

In the context of the four New World countries – Australia, New Zealand, Canada and the United States – the juxtaposition of Indigenous and non-Indigenous cultures highlights how the varying balance of these forces combines to generate radically different mobility outcomes. It also serves to underline the variant traditions of migration research among the two groups. While the chapters in this book provide mixed evidence of convergence in settlement patterns and processes, greater confluence is apparent in methods of investigation.

On the one hand there is the increasing application of mainstream quantitative methods to measure Indigenous mobility, as reflected in Part 1 of this volume. Understanding of Indigenous mobility can only benefit from the application of rigorous comparative analyses, whether against Indigenous groups in similar contextual settings, or by comparison with the majority non-Indigenous population in the same land. In this regard, current endeavours aimed at developing standard measures to capture key dimensions of migration (Bell *et al.* 2002; Rees *et al.* 2000) are as applicable to comparisons of Indigenous mobility as they are for the cross-national comparisons of mainstream migration experience for which they were originally conceived.

On the other hand it is equally apparent that the concepts, tools and techniques adapted and applied to the field of Indigenous research can also add new dimensions to understanding of non-Indigenous mobility. This is not simply to repeat the now familiar call for wider adoption of biographic and ethnographic techniques; it also points to the need for recognition of entire modes of mobility behaviour. Circulation, for example, is universally recognized as an essential strategy fulfilling multiple objectives among Indigenous communities, paralleling similar recognition in developing countries (Chapman and Prothero 1983), but has barely begun to register in research on mobility in the developed world (Bell 2001; Green *et al.* 1999).

While closer links to mainstream analysis would appear to offer synergies enriching both domains, the evidence assembled in this volume helps crystallize discrete aspects of Indigenous mobility that invite sustained attention in ongoing research. Building on the framework established in the opening to this volume, we would argue that a primary focus should be given to further elaborating the way in which mobility dynamics and settlement outcomes are shaped by the changing interface between Indigenous culture and the encapsulating state. An enhanced understanding of these interactions, in diverse settings and at varying temporal and spatial scales, is fundamental to articulation of a robust and comprehensive theory of mobility among Indigenous peoples. This, in turn, represents an essential building block in defining the enclave demographies that characterize Indigenous populations in New World settings.

Rigorous testing of such theory is crucially dependent upon the application of appropriate analytical techniques to reliable data. An essential ingredient, as canvassed in Part 2 of this volume, lies in addressing the manifold problems associated with measuring indigeneity. Equally important, and perhaps more tractable, is the task of capturing the many forms of spatial activity that characterize Indigenous life. Mobility is ultimately a 'means to ends in space', a mechanism for balancing activities across multiple life domains (Hooimeijer and van der Knaap 1994). The challenge for research is not simply to capture the dynamics of these diverse forms of movement, but to understand how they intersect and interweave to underpin the lives of Indigenous peoples.

The quest for enhanced understanding of population mobility remains one of the outstanding intellectual challenges confronting demography, but the key impetus that will drive future research for Indigenous populations lies in its relevance for policy. Given the central role of public intervention in guiding Indigenous mobility and settlement outcomes over the course of history, a policy focus seems uniquely appropriate for research in this field. The challenge for the future, however, lies not only in establishing how policies and programmes have shaped previous mobility outcomes, but in gauging the policy responses that will best facilitate the future goals and aspirations of Indigenous peoples in developed country settings.

References

Bell, M. (2001) 'Understanding circulation in Australia', *Journal of Population Research*, 18(1): 1–18.
Bell, M. Blake, M., Boyle, P., Duke-Williams, O., Hugo, G., Rees, P. and Stillwell, J. (2002) 'Cross-national comparison of internal migration: issues and measures', *Journal of the Royal Statistical Society A*, 165(3): 435–64.
Chapman, M. and Prothero, M. (1983) 'Themes on circulation in the third world', *International Migration Review*, 17: 597–632.
Green, A.E., Hogarth, T. and Schakleton, R. (1999) 'Longer distance commuting as a substitute for migration in Britain', *International Journal of Population Geography*, 5(1): 49–68.
Halfacree, K.H. and Boyle, P.J. (1993) The challenge facing migration research: the case for a biographical approach, *Progress in Human Geography*, 17(3): 333–48.
Hooimeijer, P. and Van der Knaap, G. (1994) 'From flows of people to networks of behaviour', in P. Hooimeijer, G.A. Van der Knaap, J. Van Weesep and R.I. Woods (eds) *Population Dynamics in Europe, Current Issues in Population Geography*, Netherlands Geographical Studies 173, Utrecht: Royal Netherlands Geographical Society and Department of Geography, University of Utrecht.
Ivison, D., Patton, P. and Sanders, W. (2000) *Political Theory and the Rights of Indigenous Peoples*, Melbourne: Cambridge University Press.
Nietschmann, B. (1994) 'The Fourth World: nations versus states,' in G.J. Demko

and W.B Wood (eds) *Reordering the World: Geopolitical Perspectives on the Twenty-First Century*, Boulder: Westview Press.

Rees, P., Bell, M., Duke-Williams, O. and Blake, M. (2000) 'Problems and solutions in the measurement of migration intensities: Australia and Britain compared', *Population Studies*, 54(2): 207–22.

Rowse, T. (2002) *Indigenous Futures: Choice and Development for Aboriginal and Islander Australia*, Sydney: UNSW Press.

Skeldon, R. (1990) *Population Mobility in Developing Countries: A Reinterpretation*, London: Bellhaven Press.

Skeldon, R. (1995) 'The challenge facing migration research: a case for greater awareness', *Progress in Human Geography*, 19(1): 91–6.

Skeldon, R. (1997) *Migration and Development: A Global Perspective*, London: Longman.

Zelinsky, W. (1971) 'The hypothesis of the mobility transition', *Geographical Review*, 61: 219–49.

Index

Abenaki 96
Aboriginal Administration Act (1936)
228
Aboriginal and Torres Strait Islander
Commission (ATSIC) 212, 213, 247,
253–4, 257 n12
Aboriginal Australians 13–38, 201–20,
223–36, 239–58; "all-one-family"
228; autonomy 224, 255; Ayapathu
242–3; bands 38 n3, 225;
ceremonies 17, 225–6, 230–4, 242;
citizenship 69 n8; colonial impacts
14, 15, 29, 243–6; cultural activities
234–5; discrete communities 31–2,
38 n4; education 25, 26, 27;
employment 26, 29, 208;
endosociality 235; government
policy 14, 15, 26, 29, 201–20,
253–6; Gunwinggu 18–20, 19*f*;
homeland centres (*see* outstations);
identity 14–15, 17, 33–4, 128, 224,
254–5; intermarriage 128, 236 n6;
Nyungars 228–9; "one-countrymen"
225–6; Pintupi 233; Pitjantjatjara
230–1, 232–3, 234; policy meetings
232; population 14–16, 128, 223;
Protection Boards 202; Protectors of
Aborigines 202, 244; relatedness
224, 228, 255; socio-economic status
15, 25–8, 231, 232, 244, 245;
sporting events 232; Tiwi 236 n4;
Warlpiri 232, 233; Yolngu 24–5; *see
also* Aboriginal mobility; Aboriginal
settlement patterns; walkabout
Aboriginal Family Demography Study
215–19

Aboriginal Family Voluntary
Resettlement Scheme 207
Aboriginal Land Trust 216
Aboriginal mobility 13–38; beats 21,
226–8; case studies 226–30, 240–3;
census data 122, 123; circular
mobility 16, 17–28; colonial impacts
243–6; contemporary patterns 227*f*,
230–4, 250–3; continuity and change
13–38; historical patterns 201,
240–3; Indigenous Community
Housing and Infrastructure Needs
Survey (CHINS) 23; *jilkaja* journeys
227*f*, 230, 231–4; and land use
18–20, 19*f*; lines 21; 228–9; means
of transport 228, 229, 230–4, 235,
245, 246, 254; mobility transition
16, 37–8; phases of evolution 16;
policy and the state 14, 15, 26,
201–7, 253–6, 257 n13; post-war
urbanization 16, 17, 33–7, 35*t*,
208–10, 220; push–pull factors 25–6,
26*f*; reasons for 223–4, 235–6 n2,
236 n4, 242, 250–3; regional inflows
and outflows 21, 22*f*; religious
ceremonies 17; runs 21, 228; and
services 21, 22–5, 24*f*; and social
networks 20–1; and socioeconomic
status 25–8, 27*f*, 28*f*; *see also*
Aboriginal settlement patterns; Cape
York Peninsula, Australia; walkabout
Aboriginal Peoples Survey (APS)
(Canada) 101, 129–32, 130*t*, 131*t*,
133 n2, 152, 153
Aboriginal settlement patterns 201–20,
204*f*, 239–58; Aboriginal Family

9 780415 224307